Maurice A. de Gosson

Quantum Harmonic Analysis

Advances in Analysis and Geometry

Volume 4

Maurice A. de Gosson

Quantum Harmonic Analysis

An Introduction

DE GRUYTER

Mathematics Subject Classification 2010
Primary: 32A50, 51A50, 81Qxx; Secondary: 35Q40, 35S05, 47G30

Author
Prof. Dr. Maurice A. de Gosson
University of Vienna
Faculty of Mathematics
Oskar-Morgenstern-Platz 1
1090 Vienna
Austria
maurice.de.gosson@univie.ac.at

ISBN 978-3-11-072261-1
e-ISBN (PDF) 978-3-11-072277-2
e-ISBN (EPUB) 978-3-11-072290-1
ISSN 2511-0438

Library of Congress Control Number: 2021934872

Bibliographic information published by the Deutsche Nationalbibliothek
The Deutsche Nationalbibliothek lists this publication in the Deutsche Nationalbibliografie; detailed bibliographic data are available on the Internet at http://dnb.dnb.de.

Typesetting: VTeX UAB, Lithuania
Printing and binding: CPI books GmbH, Leck

www.degruyter.com

To Charlyne and to our children
Serge, Corinne, Samantha, and Sven

With all my love

Contents

Preface —— XIII

Introduction —— XV

1 Preliminaries —— 1
1.1 Vector calculus: notation —— **1**
1.1.1 Vectors, matrices, and more —— **1**
1.1.2 Multi-index notation —— **2**
1.2 Hilbert spaces —— **3**
1.2.1 Example: the Sobolev spaces H^s —— **3**
1.2.2 The projection theorem —— **4**
1.2.3 Tensor products of Hilbert spaces —— **4**
1.2.4 Compact operators —— **5**
1.2.5 Positive operators —— **6**
1.3 Tempered distributions —— **7**
1.3.1 The Schwartz space $\mathcal{S}(\mathbb{R}^n)$ of test functions —— **7**
1.3.2 The dual space $\mathcal{S}'(\mathbb{R}^n)$ —— **7**
1.4 The Fourier transform —— **8**
1.4.1 Definition —— **8**
1.4.2 Plancherel's theorem —— **8**
1.4.3 Fourier transform and convolution —— **9**
1.5 The symplectic group —— **9**
1.5.1 Classical Lie groups —— **9**
1.5.2 Definition of $\mathrm{Sp}(n)$ —— **10**
1.5.3 Symplectic block-matrices —— **11**
1.5.4 The eigenvalues of a symplectic matrix —— **12**
1.5.5 Williamson's symplectic diagonalization —— **12**
1.6 Quantum mechanics —— **13**
1.6.1 The axioms of quantum mechanics —— **14**
1.6.2 Quantization —— **15**
1.6.3 Superposition and entanglement —— **15**
1.6.4 Mixed quantum states; the density matrix —— **16**
1.6.5 Entanglement: spooky actions at a distance —— **17**
1.6.6 Variable $\hbar$ —— **18**

2 Displacements and reflections —— 21
2.1 The Heisenberg displacement operator —— **21**
2.1.1 Definition and motivation —— **21**
2.1.2 Properties —— **22**
2.2 The Grossmann–Royer reflection operators —— **25**

2.3 The symplectic Fourier transform — 27
2.4 An analytic relationship between $\widehat{D}(z)$ and $\widehat{R}(z)$ — 28
2.5 The Schrödinger representation — 29
2.6 Comments and references — 31

3 The cross-Wigner transform — 33
3.1 Definition and properties — 33
3.1.1 Definitions — 33
3.1.2 Elementary properties — 34
3.1.3 Hudson's theorem — 36
3.1.4 Displacing cross-Wigner transforms — 36
3.2 The Moyal identity — 38
3.2.1 Statement and proof — 38
3.2.2 An extension result for the cross-Wigner transform — 39
3.2.3 A reconstruction result — 40
3.3 The cross-ambiguity function — 41
3.3.1 Definition — 41
3.3.2 Relations with the cross-Wigner transform — 41
3.3.3 Moyal identity for the cross-ambiguity function — 43
3.4 Dependence of the Wigner transform on h — 44
3.5 Statistical interpretation of the Wigner transform — 46
3.6 Comments and references — 47

4 Gaussians and hermite functions — 49
4.1 The Wigner transform of a gaussian — 49
4.1.1 The cross-Wigner transform of a pair of Gaussians — 49
4.1.2 Fermi's trick — 51
4.1.3 Cross-ambiguity function of a Gaussian — 52
4.2 The case of Hermite functions — 53
4.2.1 Hermite functions — 53
4.2.2 Laguerre polynomials and functions — 54
4.2.3 The cross-Wigner transform of Hermite functions — 55
4.3 Comments and references — 59

5 The Weyl transform — 61
5.1 Definitions of the Weyl transform — 61
5.1.1 First definition — 61
5.1.2 Definition using the Wigner transform — 63
5.1.3 Pseudo-differential definition — 65
5.2 Main properties — 66
5.2.1 The distributional kernel of a Weyl operator — 66
5.2.2 The adjoint and the transpose of a Weyl operator — 67

5.3 Shubin's symbol classes — 68
5.4 Composition formulas — 69
5.5 Comments and references — 71

6 The Cohen class — 73
6.1 Definition of the Cohen class — 73
6.2 A Moyal identity — 75
6.3 The marginal properties — 76
6.4 The τ-Wigner transform — 77
6.4.1 Definition — 77
6.4.2 Properties — 79
6.5 The Born–Jordan kernel — 81
6.6 Comments and references — 82

7 Born–Jordan quantization — 83
7.1 Physical origins — 83
7.2 Algebraic motivation — 84
7.2.1 Weyl, Shubin, and Born–Jordan — 84
7.2.2 A remark on the quantization of monomials — 85
7.3 Born–Jordan operators: definition(s) — 85
7.3.1 Definition using the Cohen class — 85
7.3.2 Pseudodifferential expression — 87
7.4 Non-invertibility of Born–Jordan quantization — 89
7.4.1 A first non-injectivity result — 90
7.4.2 A general result — 91
7.5 Comments and references — 92

8 Metaplectic operators — 93
8.1 The symplectic group $\mathrm{Sp}(n)$ — 93
8.2 The metaplectic representation — 94
8.2.1 Fourier integral operators with quadratic phases — 94
8.2.2 Definition and properties of $\mathrm{Mp}(n)$ — 96
8.3 The projection π^{Mp} — 96
8.3.1 Definition of the covering projection — 96
8.3.2 Construction of π^{Mp} — 97
8.4 Comments and references — 101

9 The property of symplectic covariance — 103
9.1 Symplectic covariance of the cross-Wigner transform — 103
9.1.1 The Heisenberg displacement operators — 103
9.1.2 The Grossmann–Royer reflections — 105
9.1.3 Symplectic covariance of $W(\psi,\phi)$ and $A(\psi,\phi)$ — 105

9.2 Application: the action of Mp(n) on Gaussians — 106
9.2.1 A first result — 106
9.2.2 Pre-Iwasawa factorization of a symplectic matrix — 107
9.2.3 Application to Gaussians — 108
9.3 Symplectic covariance of Weyl operators — 109
9.4 Maximality of the symplectic group — 111
9.4.1 A technical lemma — 111
9.4.2 The maximality property for the Wigner transform — 112
9.4.3 Maximal covariance of Weyl operators — 115
9.4.4 Born–Jordan operators — 115
9.5 Comments and references — 117

10 The Feichtinger algebra — 119
10.1 Definition and first properties — 119
10.1.1 Definition of the Feichtinger algebra — 119
10.1.2 First properties — 121
10.1.3 The Banach algebra property — 124
10.2 The metaplectic invariance of $S_0(\mathbb{R}^n)$ — 128
10.3 The dual space $S_0'(\mathbb{R}^n)$ — 129
10.4 The modulation spaces $M_s^1(\mathbb{R}^n)$ — 131
10.5 Comments and references — 133

11 Hilbert–Schmidt operators — 135
11.1 Hilbert–Schmidt operators on $L^2(\mathbb{R}^n)$ — 135
11.1.1 First definition — 135
11.1.2 Second definition — 136
11.1.3 Equivalence of the definitions; the Hilbert–Schmidt norm — 137
11.2 Further properties of the Hilbert–Schmidt operators — 140
11.2.1 Compactness of Hilbert–Schmidt operators — 140
11.2.2 Ideal property — 141
11.3 Hilbert–Schmidt and Weyl operators — 142
11.4 Comments and references — 143

12 The trace class — 145
12.1 Definitions of trace class operators — 145
12.1.1 Statement of the definitions — 145
12.1.2 Proof of the equivalence of the definitions — 146
12.1.3 Definition of the trace — 149
12.2 Properties of the trace class — 150
12.2.1 Trace class operators form an ideal — 150
12.2.2 The trace class norm — 152
12.3 Trace formulas — 153

12.3.1 A product formula — 153
12.3.2 Relationship between the trace and the Weyl symbol — 154
12.4 Comments and references — 156

13 The quantum Bochner theorem — 157
13.1 The positivity question — 157
13.2 Bochner's theorem — 157
13.3 The quantum case — 158
13.3.1 The KLM conditions — 158
13.3.2 The main result; statement and proof — 159
13.4 The case of Gaussian distributions — 161
13.4.1 Two lemmas — 161
13.4.2 The condition on Σ — 162
13.5 Positive trace-class operators — 164
13.6 Comments and references — 165

14 The density operator — 167
14.1 Mixed quantum states: heuristics — 167
14.2 Density operator: definition and first properties — 168
14.2.1 Definition of a density operator — 168
14.2.2 First properties — 168
14.2.3 The convexity property — 170
14.2.4 The spectral theorem for density operators — 171
14.3 The Wigner distribution as a Weyl symbol — 173
14.3.1 The density operator as a Weyl operator — 173
14.3.2 Application to the statistical interpretation of $\widehat{\rho}$ — 174
14.4 Quantum condition on the covariance matrix — 175
14.4.1 The covariance matrix of a density operator — 175
14.4.2 The quantum condition — 176
14.5 The Narcowich–Wigner spectrum — 179
14.6 Comments and references — 180

15 The uncertainty principle — 181
15.1 Feichtinger states — 181
15.2 The Robertson–Schrödinger inequalities — 182
15.3 Gromov's symplectic non-squeezing theorem — 184
15.3.1 Statement of Gromov's theorem — 184
15.3.2 Proof of Gromov's theorem in the linear case — 184
15.3.3 The Gromov width — 185
15.4 The covariance ellipsoid — 186
15.5 Uncertainty and quantum polarity — 188
15.5.1 Orthogonal projections of the covariance ellipsoid — 188

15.5.2 Polar duality in convex geometry — 189
15.5.3 The fundamental property of Ω_X and Ω_P — 190
15.6 Comments and references — 191

16 Separability and entanglement — 193
16.1 The reduced density operator — 193
16.1.1 Physical motivation — 193
16.1.2 Traditional Hilbert space approach — 193
16.1.3 The reduced density operator in harmonic analysis — 194
16.2 Separability and the PPT condition — 197
16.2.1 Separable and entangled quantum states — 197
16.2.2 The PPT theorem — 198
16.2.3 The Schur complement — 199
16.2.4 Orthogonal projections of the covariance ellipsoid — 200
16.3 Werner and Wolf's condition — 201
16.3.1 Statement and equivalent formulation — 201
16.3.2 A geometric consequence — 203
16.4 Comments and references — 204

17 Separability of Gaussian states — 205
17.1 The purity of a Gaussian state — 205
17.1.1 A general purity formula for Gaussian sates — 205
17.1.2 Pure Gaussian states — 206
17.2 Separability of Gaussian quantum states — 207
17.2.1 Generalities, a sufficient condition for separability — 207
17.2.2 Reduced states of a Gaussian density operator — 208
17.2.3 Back to the Werner and Wolf conditions — 210
17.3 Disentanglement of Gaussian states — 210
17.3.1 A diagonalization result for positive symplectic matrices — 210
17.3.2 The disentanglement result — 212
17.4 The case of Gaussians — 213
17.5 Comments and references — 214

Bibliography — 215

Index — 219

Preface

In the early 1970s, one of my analysis teachers Jean Dieudonné (who was, incidentally, one of the founding members of the "Bourbaki group") declared, during a visit to the *Mathematisches Kolloquium* in Vienna (organized at that time by Hans Reiter), that "harmonic analysis was off-stream" and that one ... "should no longer expect any interesting developments in this area". While Dieudonné is well-known for his provocative and unfair statements (he had also severely criticized Henri Poincaré in *A History of Algebraic and Differential Topology* for his alleged lack of rigor), he should have known better, especially, since he had himself contributed to the area! It is an ironic answer to Dieudonné that harmonic analysis has become over the years a thriving area of both pure and applied mathematics under the impetus of scientists working in several related areas, for instance, in time-frequency analysis with the study of modulation spaces initiated by Hans Feichtinger.

Quantum mechanics is arguably one of the most successful scientific theories ever; by quantum mechanics, we mean here quantum physics, quantum chemistry, or quantum optics, which refer to the same thing, even if in a few circumstances there could be a subtle difference between these topics. Historically, the story goes as follows: In the 1920's, physicists (Schrödinger, Heisenberg, Born, and many others) developed mathematical tools that described the quantum behavior of subatomic particles. At this point, physicists began calling the new field "quantum mechanics" on the model of the phrase "classical mechanics".

One could expect from such a successful scientific theory that it is based on a set of well-defined and unambiguous axioms (or postulates), as is the case for classical mechanics, which is governed by Newton's laws. Unfortunately, there seems to be no consensus among physicists about which system of axioms should be used. The difficulty (for physicists) is that quantum mechanics is still plagued with severe interpretational issues: One does not really know how to interpret the results one is dealing with (the "Wigner's friend" paradox is emblematic of these difficulties). For instance, does the wavefunction have any physical reality? And how should its collapse be interpreted? Questions like these are numerous, and one sees new and often contradictory recipes or interpretations appear everyday in specialized journals, or pre-print repositories like arXiv. Physicists have developed, with the help of philosophers, a whole industry based on the spreading of unending discussions; it has also given birth to uncountable papers or books popularizing quantum mechanics, often in weird or even wrong ways. The reader can get a taste of these issues by reading the entry

https://plato.stanford.edu/entries/qt-issues/

of the Stanford Encyclopedia of Philosophy. As the physicist Richard Feynman put it, somewhat jokingly,

> "Philosophy of science is about as useful to scientists as ornithology is to birds."

https://doi.org/10.1515/9783110722772-201

For a mathematician, this situation is of course not tenable, and we will view quantum mechanics in this book as a branch of mathematics based on a well-defined and unambiguous system of axioms. There are many possible choices of (equivalent, or non-equivalent) such systems. We will apply the principle of parsimony and propose the simplest (and therefore the most general) set of axioms.

It is my pleasure and duty to thank the following mathematicians and friends for fruitful conversations and encouragements. In alphabetical and geographical order: Nuno Dias and Joao Prata (Lisbon), Basil Hiley and Glen Dennis (London), Leon Cohen (New York), Jean-Pierre Gazeau (Paris), Franz Luef (Trondheim), Paolo Boggiatto, Elena Cordero, Fabio Nicola, Luigi Rodino (Turin), Markus Faulhuber, Hans Feichtinger, and Karlheinz Gröchenig (Vienna). My deepest gratitude goes to Charlyne de Gosson for having read the manuscript and pointed out several typos and errors.

Acknowledgement

This work has been financed by the Grant number P 33447 of the Austrian Research Agency FWF (Fonds zur Förderung der wissenschaftlichen Forschung).

Introduction

This book is about mathematical techniques relevant to the study of theoretical quantum mechanics. As its title indicates, it is an *introduction* to the vast topic of quantum harmonic analysis, and *not* a treatise on general harmonic analysis. We therefore do not aim at completeness; for reasons of time and space, several topics have been omitted; for instance, we do not discuss the important notions of quantum channels and completely positive maps, nor do we give physical applications (this is after all a book on mathematics, not physics!). However, we discuss somewhat in depth some difficult subjects that are still being very much investigated in current research; for instance, the questions of positivity for density operators and the separability and entanglement of quantum states, which are notoriously difficult. These are advanced notions where open mathematical problems still abound.

The book is structured as a series of lecture notes, consisting of rather short chapters each of which can be – ideally – taught during a 90-minute session. The prerequisites are rather modest: The reader is supposed to have a basic knowledge of linear algebra and analysis (advanced calculus). The basic notions of functional analysis (mainly the theory of Hilbert spaces and distributions) are reviewed in a preliminary chapter, mostly without proofs. Some familiarity with elementary quantum mechanics would certainly be helpful, but is not required.

More precisely:

- Chapter 1 is a review of elementary properties from analysis and geometry. This chapter is meant to be a reference tool for the reader; we recall the basic notions from multivariate calculus, function spaces, and distribution theory, while fixing the notation we will use. If a reader thinks he is sufficiently familiar with these topics, he is encouraged to skip this chapter and directly proceed to the core of the book, beginning in next chapter.
- Chapters 2 and 3 set the foundations for harmonic analysis in quantum mechanics: We introduce two elementary mathematical objects, the Heisenberg displacement operator and the Grossmann–Royer reflection operator, and use them to construct the Wigner transform and its symplectic Fourier transform, the ambiguity function. This approach to the Wigner formalism is not quite conventional, but has many advantages (conceptual and practical). Perhaps one of the most useful formulas that are derived in Chapter 2 is the Moyal identity. It is — among many other things — a bridge between orthogonality in $L^2(\mathbb{R}^n)$ and orthogonality in $L^2(\mathbb{R}^{2n})$.
- In Chapter 4, we take a short break from theory and use the Wigner formalism previously introduced to calculate explicitly the cross-Wigner transforms of Gaussian and Hermite functions, thus obtaining a useful collection of formulas that are otherwise found spread about in the literature. It gives us the opportunity to quickly review some notions from the theory of special functions and polynomials.

https://doi.org/10.1515/9783110722772-202

- In Chapter 5, we are back to work. We define and study the properties of the arguably most important "quantization" procedure, the Weyl transform. Its genuine mathematical interest (besides being a theory of pseudodifferential operators) come from Schwartz's distributional kernel theorem, which allows to prove that every linear continuous operator, from the space of Schwarz test functions to the space of tempered destitutions, is a Weyl operator, in the sense that one can associate to it (in a unique way) a symbol, which is a function (or tempered distribution) defined on phase space. (In the language of quantum mechanics, one would say that to every Weyl operator corresponds a unique "classical observable", and *vice versa*).
- Chapter 6 is about the so-called Cohen class. It is the key to the alternative quantization scheme due to Born and Jordan we study in next chapter. Elements of the Cohen class are the convolutions of the Wigner transform with adequate temperate distributions ("Cohen kernels"). They are generalizations of the Wigner transform, and one recovers the latter by choosing as Cohen kernel the Dirac measure. The importance of the Cohen class has been growing over recent years, especially in time-frequency analysis and the study and damping of interference effects. In our case, it will be instrumental for the definition of Born–Jordan quantization as an alternative to the Weyl transform.
- In Chapter 7, which is devoted to a quantization procedure originally due (in a physical form) to Max Born and Pascual Jordan (and Werner Heisenberg), we begin by discussing (as an appetizer) some ordering issues for the quantization of monomials, and thereafter define the Born–Jordan quantization of arbitrary symbols. Born–Jordan quantization might very well be the physically correct way of associating operators to classical observables in quantum theory. It is constructed using an averaging procedure involving Shubin's τ-pseudodifferential operators.
- Chapter 8 is devoted to the metaplectic machinery, which is in principle well-known (at least in its rudimentary forms) to the harmonic analysis community. Quite abstractly, the metaplectic group is a unitary representation on the square-integrable functions of the double cover of the symplectic group. Its importance comes from the fact that to every symplectic matrix the metaplectic representation associates two unitary operators, and these operators represent in a sense quantizations of the corresponding symplectic matrices.
- In Chapter 9, we use the metaplectic representation of the symplectic group to study the so-called symplectic covariance property of Weyl quantization. This is an essential property, characteristic of the Weyl transform, and is extremely useful in all applications since it allows to considerably simplify many problems. Roughly speaking, symplectic covariance means that, if one makes a linear symplectic change of variables S in a symbol, then the corresponding Weyl operator is conjugated by anyone of the metaplectic operators $\pm\widehat{S}$ arising from that change of variables.

- Chapter 10 is devoted to a short study of the so-called Feichtinger algebra and to its most elementary weighted extension. The Feichtinger algebra is the simplest of all modulation spaces, a class of functional spaces introduced in the mid-1980s by Hans Feichtinger. While they are usually defined using the Gabor transform, in our approach the properties of he Wigner transform are used; this not only makes their theory explicitly symplectically covariant, but is also much better adapted to the study of quantum mechanical objects (for instance, the density operator). The importance of the Feichtinger algebra and its extensions comes in our context from the fact that they provide us with the right framework for studying density operators and the associated mixed quantum states.
- Chapters 11 and 12 are about two very well-known and related classes of operators, the Hilbert–Schmidt and trace-class operators. In addition to their intrinsic interest, these chapters are necessary preparation for a rigorous mathematical study of the density operator of quantum mechanics.
- Chapter 13 is a technical chapter. It is about a rather difficult topic, that of the positivity of trace-class operators. Positivity questions have been of great concern since the very beginning of the theory of pseudodifferential operators, and many problems are still open. We give a new proof of the so-called quantum Bochner theorem, which is the most important known necessary and sufficient condition for the positivity of a trace-class operator.
- Chapter 14 is about mixed states and the corresponding density operator. A mixed quantum state is the datum of a set of pairs $\{(\psi_j, \lambda_j)\}$ where ψ_j is a (normalized square integrable) pure state and λ_j a classical probability; these probabilities sum up to one: $\sum_j \lambda_j = 1$. The study of these objects is at the heart of quantum mechanics, both for theoretical and practical reasons. One could say that quantum mechanics ultimately is the study of density operators.
- In Chapter 15, we analyze the relationship between the positivity properties of density operators and the quantum uncertainty principle from a rigorous point of view, where the usual Heisenberg and Robertson–Schrödinger inequalities play only an accessory role. We also propose a new approach to quantum uncertainty using convex geometry (the property of polar duality).
- Chapters 16 and 17 are devoted to the study of the notions of entanglement and separability of quantum states. These notions play a very privileged role in quantum mechanics and are at the heart of modern research. While not so much is known about the general characterization of entanglement and separability of arbitrary mixed states, the situation for Gaussian states is somewhat more satisfactory. Not only are these states somewhat easier to study analytically than general states, but they are also omnipresent in many applications (quantum optics and computing, for instance).

1 Preliminaries

We collect in this introductory chapter some basic reference material about Hilbert spaces, distribution spaces, and symplectic geometry. The properties are given mostly without proof (an exception being the proof of Williamson's symplectic diagonalization theorem). For a detailed study of the symplectic group and its properties (and, more generally, of symplectic geometry), we refer to our book [30]; for functional analysis at an elementary level [6, 57] are good readings.

1.1 Vector calculus: notation

1.1.1 Vectors, matrices, and more

In matrix calculations, elements of $\mathbb{R}^n$ should be viewed as column vectors; for instance

$$x = \begin{pmatrix} x_1 \\ \vdots \\ x_n \end{pmatrix}.$$

However, for typographic economy, we will usually write $x = (x_1, \ldots, x_n)$ in the text. The Euclidean scalar product $\cdot$ and norm $|\cdot|$ on $\mathbb{R}^n$ are defined by

$$x \cdot y = x^T y = \sum_{j=1}^{n} x_j y_j, \quad |x| = \sqrt{x \cdot x}.$$

We will often write $x \cdot x = x^2$ when no confusion is likely to arise. The gradient operator in the variables $x_1, \ldots, x_n$ is

$$\partial_x = \begin{pmatrix} \frac{\partial}{\partial x_1} \\ \vdots \\ \frac{\partial}{\partial x_n} \end{pmatrix},$$

or, with the same abuse of notation as above, $\partial_x = (\partial/\partial x_1, \ldots, \partial/\partial x_n)$. Let f and g be differentiable functions $\mathbb{R}^n \longrightarrow \mathbb{R}^n$; in matrix form, the chain rule is

$$\partial(g \circ f)(x) = (Df(x))^T \partial f(x) \tag{1.1}$$

where $Df(x)$ is the Jacobian matrix of f: If $f = (f_1, \ldots, f_m)$ is a differentiable mapping $\mathbb{R}^m \longrightarrow \mathbb{R}^m$ then

$$Df = \begin{pmatrix} \frac{\partial f_1}{\partial x_1} & \frac{\partial f_1}{\partial x_2} & \cdots & \frac{\partial f_1}{\partial x_m} \\ \frac{\partial f_2}{\partial x_1} & \frac{\partial f_2}{\partial x_2} & \cdots & \frac{\partial f_2}{\partial x_m} \\ \vdots & \vdots & \ddots & \vdots \\ \frac{\partial f_m}{\partial x_1} & \frac{\partial f_n}{\partial x_2} & \cdots & \frac{\partial f_m}{\partial x_m} \end{pmatrix}. \tag{1.2}$$

https://doi.org/10.1515/9783110722772-001

Let $y = f(x)$; we will indifferently use the notations

$$Df(x),\ \frac{\partial y}{\partial x},\ \frac{\partial(y_1,\dots,y_m)}{\partial(x_1,\dots,x_m)}$$

for the Jacobian matrix. If f is invertible, the inverse function theorem says that

$$D(f^{-1})(y) = \left[Df(x)\right]^{-1}. \tag{1.3}$$

If $f : \mathbb{R}^m \longrightarrow \mathbb{R}$ is a twice continuously differentiable function, its Hessian calculated at a point x is the symmetric matrix of second derivatives

$$D^2 f(x) = \begin{pmatrix} \frac{\partial^2 f}{\partial x_1^2} & \frac{\partial^2 f}{\partial x_1 \partial x_2} & \cdots & \frac{\partial^2 f}{\partial x_1 \partial x_n} \\ \frac{\partial^2 f}{\partial x_2 \partial x_1} & \frac{\partial^2 f}{\partial x_2^2} & \cdots & \frac{\partial^2 f}{\partial x_2 \partial x_n} \\ \vdots & \vdots & \ddots & \vdots \\ \frac{\partial^2 f}{\partial x_n \partial x_1} & \frac{\partial^2 f}{\partial x_n \partial x_2} & \cdots & \frac{\partial^2 f}{\partial x_n^2} \end{pmatrix}. \tag{1.4}$$

Notice that the Jacobian and Hessian matrices are related by the formula

$$D(\partial f)(x) = D^2 f(x). \tag{1.5}$$

Also note the following useful formulae:

$$A\partial_x \cdot \partial_x e^{-\frac{1}{2}Mx\cdot x} = (MAMx \cdot x - \mathrm{Tr}(AM))e^{-\frac{1}{2}Mx\cdot x} \tag{1.6}$$

$$Bx \cdot \partial_x e^{-\frac{1}{2}Mx\cdot x} = (MBx \cdot x)e^{-\frac{1}{2}Mx\cdot x} \tag{1.7}$$

where A, B, and M are symmetric $n \times n$ matrices.

1.1.2 Multi-index notation

By convention, the set of all non-negative integers is denoted by $\mathbb{N}$:

$$\mathbb{N} = \{0, 1, 2, \dots\}$$

(this convention is not universal; in some countries 0 is not an element of $\mathbb{N}$). By definition, a (n-)multi-index is an element $\alpha = (\alpha_1, \dots, \alpha_n)$ of $\mathbb{N}^n$. We call the integer $|\alpha| = \alpha_1 + \cdots + \alpha_m$ the "length" of the "multi-index" α and define, for $x \in \mathbb{R}^n$, the "power " x^α by

$$x^\alpha = x_1^{\alpha_1} \cdots x_n^{\alpha_n}, \quad \partial_x^\alpha = \partial_{x_1}^{\alpha_1} \cdots \partial_{x_n}^{\alpha_n}.$$

Likewise,

$$\alpha! = \alpha_1! \cdots \alpha_n!, \quad \binom{\alpha}{\beta} = \frac{\alpha!}{\beta!(\alpha - \beta)!}$$

when $\alpha \geq \beta$, that is when $\alpha_j \geq \beta_j$ for $1 \leq j \leq n$.

1.2 Hilbert spaces

Hilbert spaces (and their cousins, Banach spaces) are the best-known functional spaces in elementary functional analysis. By definition, a complex Hilbert space is a vector space H having the following properties:

- H is endowed with an inner product (also called a scalar product) $(\cdot,\cdot)_H$, that is a sesquilinear form $(\cdot,\cdot)_H : H \times H \longrightarrow \mathbb{C}$ such that (i) $(\psi,\psi)_H > 0$ for all $\psi \in H$, $\psi \neq 0$; (ii) $(\psi,\phi)_H = 0$ for all $\phi \in H$ implies that $\psi = 0$; (iii) $(\psi,\phi)_H = \overline{(\phi,\psi)_H}$. We suppose, by convention, that the anti-linearity takes place in the second term of the scalar product, i. e., $(\lambda\psi,\phi) = \lambda(\psi,\phi)$ and $(\psi,\lambda\phi) = \overline{\lambda}(\psi,\phi)$ for $\lambda \in \mathbb{C}$;
- Let $\|\cdot\|_H$ be the norm associated with $(\cdot,\cdot)_H$: $\|\psi\|_H = (\psi,\psi)_H^{1/2}$. The space H is complete for the topology associate with this norm; i. e., every Cauchy sequence $(\psi_j)_j$ is convergent.

The standard example (and perhaps the most useful in harmonic analysis and quantum mechanics) is the space $L^2(\mathbb{R}^n)$ of all square-integrable functions equipped with the scalar product

$$(\psi|\phi)_{L^2} = \int_{\mathbb{R}^n} \psi(x)\overline{\phi(x)}dx.$$

We mention that the scalar product $(\psi|\phi)_{L^2}$ is related to the physicist's "bra-ket" product $\langle\psi|\phi\rangle$ by conjugation:

$$\langle\psi|\phi\rangle = \overline{(\psi|\phi)_{L^2}}$$

which reflects the fact that physicists mostly use the opposite convention for sesquilinearity of the scalar product of functions.

1.2.1 Example: the Sobolev spaces H^s

The Sobolev spaces H^s are well-known functional spaces, much used in the theory of partial differential operators and in microlocal analysis.

Definition 1. Let $s \in \mathbb{R}$. The Sobolev space $H^s(\mathbb{R}^n)$ consists of all $\psi \in \mathcal{S}'(\mathbb{R}^n)$ whose Fourier transforms $\widehat{\psi} = F\psi$ are functions satisfying

$$\|\psi\|_{H^s} = \Big(\int_{\mathbb{R}^n} |\widehat{\psi}(p)|^2(1+|p|^2)^s dp\Big)^{1/2} < \infty.$$

In particular, $H^0(\mathbb{R}^n) = L^2(\mathbb{R}^n)$, and we have $H^s(\mathbb{R}^n) \subset H^{s'}(\mathbb{R}^n)$ if $s' \leq s$. It is clear that $H^s(\mathbb{R}^n)$ is a vector space, and one proves that $\psi \longmapsto \|\psi\|_{H^s}$ is a norm on $H^s(\mathbb{R}^n)$

that endows it with a Banach space topology. Moreover, one has the inclusions

$$H^s(\mathbb{R}^n) \subset C^k(\mathbb{R}^n) \text{ for } s > \frac{n}{2} + k$$

and

$$\bigcap_{s\geq 0} H^s(\mathbb{R}^n) = \mathcal{S}(\mathbb{R}^n).$$

We also have $\delta \in H^s(\mathbb{R}^n)$ for $s < -n/2$, and, in fact, $H^s(\mathbb{R}^n)$ is a Hilbert space for the scalar product

$$(\psi|\phi)_{H^s} = \int_{\mathbb{R}^n} \widehat{\psi}(p)\overline{\widehat{\phi}(p)}(1 + |p|^2)^s dp.$$

1.2.2 The projection theorem

The main virtue of Hilbert spaces in quantum mechanics is that one can define unambiguously orthogonal projections on convex subsets of that space (hence, in particular, on linear subspaces). This property is summarized in the following theorem

Theorem 2. *Let C be a non-empty convex subset of the Hilbert space H. For every $\psi \in H$, there exists a unique $\psi_0 \in C$ such that*

$$\|\psi - \psi_0\|_H \leq \|\psi - \phi\|_H \quad \text{for all } \phi \in C$$

that is $\|\psi - \psi_0\| \leq d(\psi, C)$ (d the distance function). The vector ψ_0 is the only element of C satisfying

$$\operatorname{Re}(\psi - \psi_0|\phi - \psi_0)_H \leq 0 \quad \text{for all } \phi \in C.$$

Let $(\psi_j)_j$ be an orthonormal basis of H: $(\psi_j|\psi_k)_H = \delta_{jk}$. Then the Bessel identity holds:

$$\|\psi\|_H^2 = \sum_j |(\psi|\psi_j)_H|^2 \tag{1.8}$$

for every $\psi \in H$.

This property is the main motivation for the use of Hilbert spaces in quantum mechanics because it allows analyzing the reduction of wavefunctions.

1.2.3 Tensor products of Hilbert spaces

Let H_1 and H_2 be two Hilbert spaces and $(\psi_1, \psi_2) \in H_1 \times H_2$. The tensor product $\psi_1 \otimes \psi_2$ is the form $H_1 \times H_2 \longrightarrow \mathbb{C}$ defined by

$$(\psi_1 \otimes \psi_2)(\phi_1, \phi_2) = (\psi_1|\phi_1)_{H_1}(\psi_2|\phi_2)_{H_2}.$$

Let F be the set of all finite linear combinations $f = \sum_{i=1}^{k} \lambda_i(\psi_{1,i} \otimes \psi_{2,i})$ of such forms. It has a natural vector space structure. We equip F with the inner product

$$\Big(\sum_i \lambda_i(\psi_{1,i} \otimes \psi_{2,i}) \Big| \sum_j \lambda'_j(\psi'_{1,j} \otimes \psi'_{2,j})\Big)_F = \sum_{i,j} \lambda_i \overline{\lambda'_j}(\psi_{1,i}|\psi'_{1,j})(\psi_{2,i}|\psi'_{2,j}).$$

It is easy to show that this definition does not depend on the choice of linear combinations representing the elements of F and that we have $(f|f) > 0$ for every $f \in F$.

Definition 3. The tensor product $H_1 \otimes H_2$ of the two Hilbert spaces H_1 and H_2 is the completion of the vector space F equipped with the inner product $(\cdot|\cdot)_F$.

Associated with the tensor product of Hilbert spaces is the tensor product of operators. We limit ourselves here to the case of bounded operators $A \in \mathcal{B}(H_1)$ and $B \in \mathcal{B}(H_2)$.

Definition 4. The tensor product $A \otimes B$ is the bounded operator on $H_1 \otimes H_2$ defined by

$$(A \otimes B)\Big(\sum_i \lambda_i \psi_{1,i} \otimes \psi_{2,i}\Big) = \sum_i \lambda_i A(\psi_{1,i} \otimes B\psi_{2,i}).$$

1.2.4 Compact operators

We denote by $\mathcal{B}(H)$ the vector space of all bounded linear operators $\widehat{A} : \mathcal{B}(H) \longrightarrow \mathcal{B}(H)$. We have $\widehat{A} \in \mathcal{B}(H)$ if and only if $\widehat{A}$ is continuous at 0, which is equivalent to saying that $\|\widehat{A}\psi\|_H \leq \|\widehat{A}\| \, \|\psi\|_H$ where

$$\|\widehat{A}\| = \sup\{\|\widehat{A}\psi\|_H : \|\psi\|_H \leq 1\}.$$

Definition 5. An operator $\widehat{A} \in \mathcal{B}(H)$ is said to be compact if, whenever (ψ_j) is a bounded sequence in H, then $(\widehat{A}\psi_j)$ contains a convergent subsequence. Equivalently, $\widehat{A}$ is compact if and only if it takes every bounded subset of H to a relatively compact subset of H.

We assumed in the definition from the beginning that a compact is bounded, so this condition is actually redundant because every operator on a Hilbert space satisfying the conditions above is automatically compact. Compact operators form a subspace $\mathcal{K}(H)$ of $\mathcal{B}(H)$ and that subspace is closed in $\mathcal{B}(H)$: Assume that $(\widehat{A}_j)$ is a sequence in $\mathcal{K}(H)$ such that there exists $\widehat{A} \in \mathcal{B}(H)$ with $\lim_{j\to\infty} \|\widehat{A}_j - \widehat{A}\| = 0$. Then $\widehat{A} \in \mathcal{K}(H)$. It immediately follows from the definition that $\widehat{A} \in \mathcal{K}(H)$ if and only if $\widehat{A}^* \in \mathcal{K}(H)$. Also:

Proposition 6. *Assume that H is separable. Every $\widehat{A} \in \mathcal{K}(H)$ is the limit (in the operator norm) of a sequence of finite rank operators.*

(Recall that an operator $\widehat{A}$ on H is of finite rank if and only if the range of $\widehat{A}$ is finite-dimensional).

Of particular interest are the spectral properties of compact self-adjoint operators. The following theorem is essential:

Theorem 7 (Spectral Theorem). *Assume that $\widehat{A} \in \mathcal{K}(H)$ is self-adjoint: $\widehat{A} = \widehat{A}^*$. Then:* (i) *The set (λ_j) of eigenvalues of $\widehat{A}$ consists of real numbers, and is at most countable, and if infinite $\lim_{j\to\infty} \lambda_j = 0$;* (ii) *The multiplicity of every eigenvalue λ_j is finite;* (iii) *If $\widehat{A} \neq 0$, there exists a most countable set (ψ_j) of orthonormal eigenvectors of $\widehat{A}$ with eigenvalues λ_j such that, if ψ_j is an eigenvector for λ_j, then*

$$\widehat{A}\psi = \sum_j \lambda_j (\psi|\psi_j)_H \psi_j. \tag{1.9}$$

The spectral theorem for compact operators plays a fundamental role in the study of the density operator in quantum mechanics.

1.2.5 Positive operators

An operator $\widehat{A} \in \mathcal{B}(H)$ $(\widehat{A} \geq 0)$ is said to be positive semidefinite if we have $(\widehat{A}\psi|\psi)_H \geq 0$ for every $\psi \in H$. Assuming that H is a complex Hilbert space, we have:

Proposition 8. *Let $\widehat{A} \in \mathcal{B}(H)$; we have $\widehat{A} \geq 0$ if and only if any of the two equivalent statements hold:* (i) *There exists $\widehat{B} \in \mathcal{B}(H)$ such that $\widehat{A} = \widehat{B}^*\widehat{B}$;* (ii) *There exists a unique $\widehat{C} \in \mathcal{B}(H)$, $\widehat{C} \geq 0$, such that $\widehat{A} = \widehat{C}^2$.*

In particular, a positive semidefinite bounded operator on a complex Hilbert space is self-adjoint. The operator $\widehat{C}$ defined by (ii) is called the *square* root of $\widehat{A}$ and denoted by $\widehat{A}^{1/2}$ (or $\sqrt{\widehat{A}}$). By definition, the operator $|\widehat{A}| = (\widehat{A}^*\widehat{A})^{1/2}$ is the modulus (or absolute value) of $\widehat{A}$.

Proposition 9. *Every self-adjoint $\widehat{A} \in \mathcal{B}(H)$ can be written (in a unique way) as $\widehat{A} = \widehat{A}_+ - \widehat{A}_-$ where $\widehat{A}_+ \geq 0$ and $\widehat{A}_- \geq 0$ and belongs to $\mathcal{B}(H)$.*

When $\widehat{A}$ is in addition compact, one way to prove this is to use the spectral formula (1.9): then

$$\widehat{A}_+\psi = \sum_j \lambda_j^+ (\psi|\psi_j)_H \psi_j, \quad \widehat{A}_-\psi = \sum_j \lambda_j^- (\psi|\psi_j)_H \psi_j$$

where the λ_j^+ (resp. λ_j^-) are the positive (*resp.* negative) eigenvalues of $\widehat{A}$.

1.3 Tempered distributions

The vector space of k times continuously differential functions $\mathbb{R}^m \longrightarrow \mathbb{C}$ is denoted by $C^k(\mathbb{R}^m)$; k is here an integer ≥ 1 or ∞. The subspace of $C^k(\mathbb{R}^m)$ consisting of the compactly supported functions is denoted by $C_0^k(\mathbb{R}^m)$.

1.3.1 The Schwartz space $\mathcal{S}(\mathbb{R}^n)$ of test functions

$\mathcal{S}(\mathbb{R}^n)$ is the Schwartz space of rapidly decreasing functions: $\psi \in \mathcal{S}(\mathbb{R}^n)$ if and only if, for every pair (α, β) of multi-indices, there $K_{\alpha\beta} > 0$ such that

$$|x^\alpha \partial_x^\beta \psi(x)| \leq K_{\alpha\beta} \quad \text{for all } x \in \mathbb{R}^n. \tag{1.10}$$

In particular, every C^∞ function on $\mathbb{R}^n$ vanishing outside a bounded set is in $\mathcal{S}(\mathbb{R}^n)$: $C_0^\infty(\mathbb{R}^n) \subset \mathcal{S}(\mathbb{R}^n)$. Taking the best constants $K_{\alpha\beta}$ in (1.10), we obtain a family of seminorms on $\mathcal{S}(\mathbb{R}^n)$

$$\|\psi\|'_{\alpha,\beta} = \sup_{x\in\mathbb{R}^n} |x^\alpha \partial_x^\beta \psi(x)|$$

and one shows that $\mathcal{S}(\mathbb{R}^n)$ is a Fréchet space for the topology defined by these seminorms. We have continuous inclusions

$$C_0^\infty(\mathbb{R}^n) \subset \mathcal{S}(\mathbb{R}^n) \subset L^2(\mathbb{R}^n).$$

Using the generalized Leibniz formula, one shows that an equivalent system of seminorms on $\mathcal{S}(\mathbb{R}^m)$ is given by

$$\|\psi\|'_{\alpha,\beta} = \sup_{x\in\mathbb{R}^n} |\partial_x^\alpha x^\beta \psi(x)|. \tag{1.11}$$

1.3.2 The dual space $\mathcal{S}'(\mathbb{R}^n)$

The topological dual of $\mathcal{S}(\mathbb{R}^n)$ is called the *space of tempered distributions* and is denoted by $\mathcal{S}'(\mathbb{R}^n)$. The distributional pairing between $\psi \in \mathcal{S}'(\mathbb{R}^n)$ and $\phi \in \mathcal{S}(\mathbb{R}^n)$ is denoted by $\langle\psi, \phi\rangle$. When ψ and ϕ are both in $\mathcal{S}(\mathbb{R}^n)$, then

$$\langle\psi, \phi\rangle = \int_{\mathbb{R}^n} \psi(x)\phi(x)dx.$$

The natural inclusions

$$\mathcal{S}(\mathbb{R}^n) \subset L^2(\mathbb{R}^n) \subset \mathcal{S}'(\mathbb{R}^n)$$

are continuous, and $\mathcal{S}(\mathbb{R}^n)$ is dense in $L^2(\mathbb{R}^n)$.

Every polynomial in the x variables defines a tempered distribution. If $P(x)$ is such a polynomial and $\psi \in \mathcal{S}'(\mathbb{R}^n)$, then $P(x)\psi \in \mathcal{S}'(\mathbb{R}^n)$ and is defined by $\langle P\psi, \phi\rangle = \langle \psi, P\phi\rangle$. More generally, P can be replaced with any C^∞ function of x such that $|\partial_x^\alpha P(x)|$ is dominated by a polynomial for every $\alpha \in \mathbb{N}^n$. The space $\mathcal{S}'(\mathbb{R}^n)$ is invariant by Fourier transform: If $\psi \in \mathcal{S}'(\mathbb{R}^n)$, then $F\psi \in \mathcal{S}'(\mathbb{R}^n)$ is defined by $\langle F\psi, \phi\rangle = \langle \psi, F\phi\rangle$.

1.4 The Fourier transform

1.4.1 Definition

Let $\psi \in L^1(\mathbb{R}^n)$. By definition, the Fourier transform of ψ is the function defined by the absolutely convergent integral

$$F\psi(p) = \widehat{\psi}(p) = \left(\frac{1}{2\pi\hbar}\right)^{n/2} \int_{\mathbb{R}^n} e^{-\frac{i}{\hbar}p\cdot x}\psi(x)dx\,;$$

here $\hbar$ is any positive number (it plays the role of a parameter and is identified in physics with $h/2\pi$ where h is Planck's constant). We have $F\psi \in C^0(\mathbb{R}^n)$ and $\lim_{|p|\to\infty} F\psi = 0$ (Riemann–Lebesgue lemma); in fact, $F\psi$ is uniformly continuous. For $\psi \in \mathcal{S}(\mathbb{R}^n)$, the following formulas hold:

$$F(x^\alpha\psi) = (i\hbar\partial_p)^\alpha F\psi, \quad F((-i\hbar\partial_x)^\alpha\psi) = p^\alpha F\psi.$$

Assume that both $\psi \in L^1(\mathbb{R}^n)$ and $\widehat{\psi} \in L^1(\mathbb{R}^n)$; then the Fourier inversion formula holds:

$$\psi(x) = \left(\frac{1}{2\pi\hbar}\right)^{n/2} \int_{\mathbb{R}^n} e^{\frac{i}{\hbar}p\cdot x}\widehat{\psi}(p)dp.$$

It follows that F is an automorphism $\mathcal{S}(\mathbb{R}^n) \longrightarrow \mathcal{S}(\mathbb{R}^n)$ that can be extended into an automorphism $\mathcal{S}'(\mathbb{R}^n) \longrightarrow \mathcal{S}'(\mathbb{R}^n)$ by defining the extension of F, also denoted by F, by duality:

$$\langle F\psi, \phi\rangle = \langle \psi, F\phi\rangle$$

for $\psi \in \mathcal{S}'(\mathbb{R}^n)$ and $\phi \in \mathcal{S}(\mathbb{R}^n)$.

1.4.2 Plancherel's theorem

An essential property of the Fourier transform is the Plancherel theorem: If $\psi \in L^1(\mathbb{R}^n) \cap L^2(\mathbb{R}^n)$, then

$$\|F\psi\|_{L^2} = \|\psi\|_{L^2}.$$

Plancherel's formula allows extending the Fourier transform to a unitary operator on $L^2(\mathbb{R}^n)$:

$$(F\psi|F\phi)_{L^2} = (\psi|\phi)_{L^2}.$$

Note that, if $\psi \in L^2(\mathbb{R}^n)$ and $\psi \notin L^1(\mathbb{R}^n)$, then $F\psi$ cannot be calculated pointwise, and one has to use some approximation procedure. One can proceed as follows (see K. Gröchenig's book *Foundations of Time Frequency Analysis*): since $L^1(\mathbb{R}^n) \cap L^2(\mathbb{R}^n)$ is dense in $L^2(\mathbb{R}^n)$, we may choose a sequence $(\psi_j)_j$ in $L^1(\mathbb{R}^n) \cap L^2(\mathbb{R}^n)$ such that $\lim_{j\to\infty} \|\psi - \psi_j\|_{L^2} = 0$. Then $F\psi_j$ is well defined, and by Plancherel's theorem $\|F\psi - F\psi_j\|_{L^2} = \|\psi - \psi_j\|_{L^2}$ so that $(F\psi_j)_j$ is a Cauchy sequence in $L^2(\mathbb{R}^n)$. The latter being complete, the sequence $(F\psi_j)_j$ has a unique limit in $L^2(\mathbb{R}^n)$. We then define the Fourier transform of $\psi \in L^2(\mathbb{R}^n)$ by $F\psi = \lim_{j\to\infty} F\psi_j$.

1.4.3 Fourier transform and convolution

The Fourier transform F (and its inverse F^{-1}) takes, under suitable conditions, convolutions to ordinary products (up to a constant factor):

$$\begin{aligned} F(\psi * \phi) &= (2\pi\hbar)^{n/2} F\psi F\phi \\ F(\psi\phi) &= (2\pi\hbar)^{-n/2} F\psi * F\phi. \end{aligned}$$

These formulas make sense if, for instance, ψ and ϕ are in $\mathcal{S}(\mathbb{R}^n)$. They also are verified if $\psi \in L^1(\mathbb{R}^n)$ and $\phi \in L^1(\mathbb{R}^n)$ since $L^1(\mathbb{R}^n)$ is a convolution algebra.

1.5 The symplectic group

1.5.1 Classical Lie groups

In what follows, $\mathbb{K} = \mathbb{R}$ or $\mathbb{C}$.

$M(m, \mathbb{K})$ is the algebra of all $m \times m$ matrices with entries in $\mathbb{K}$.

$\mathrm{GL}(m, \mathbb{K})$ is the general linear group. It consists of all invertible matrices in $M(m, \mathbb{K})$.

$\mathrm{SL}(m, \mathbb{K})$ is the special linear group: It is the subgroup of $\mathrm{GL}(m, \mathbb{K})$ consisting of all the matrices with determinants equal to one.

$\mathrm{Sym}(m, \mathbb{K})$ is the vector space of all symmetric matrices in $M(m, \mathbb{K})$; it has dimension $\frac{1}{2}m(m+1)$. $\mathrm{U}(n, \mathbb{C})$ is the unitary group; it consists of all $U \in M(n, \mathbb{C})$ such that $UU^* = U^*U = I$ ($U^* = \bar{U}^T$ is the adjoint of U).

$\mathrm{U}(n, \mathbb{C})$ is the multiplicative group of all unitary isomorphisms of $\mathbb{R}^n$: $u \in \mathrm{U}(n, \mathbb{C})$ if and only $u^*u = uu^* = I_{n\times n}$.

$U(n)$ is the image in $GL(2n, \mathbb{R})$ of $U(n, \mathbb{C})$ by the monomorphism

$$A + iB \longmapsto \begin{pmatrix} A & -B \\ B & A \end{pmatrix} \tag{1.12}$$

where A and B are real $n \times n$ matrices.

1.5.2 Definition of Sp(n)

The symplectic group $\mathrm{Sp}(n)$ deserves a special mention since it will be ubiquitous in this book. $\mathrm{Sp}(n)$ is the group of all (linear) automorphisms of $\mathbb{R}^{2n} \equiv \mathbb{R}^n \times \mathbb{R}^n$ leaving the symplectic form

$$\sigma(z, z') = \sum_{j=1}^{n} p_j x_j' - p_j' x$$

invariant:

$$S \in \mathrm{Sp}(n) \Longleftrightarrow \sigma(Sz, Sz') = \sigma(z, z')$$

for all $(z, z') \in \mathbb{R}^{2n} \times \mathbb{R}^{2n}$. Choosing once and for all a symplectic basis (for instance, the canonical basis of $\mathbb{R}^n \times \mathbb{R}^n$), $\mathrm{Sp}(n)$ can be identified with a matrix group, in fact, with the subgroup of $GL(2n, \mathbb{R})$ consisting of all real $2n \times 2n$ matrices S such that

$$S^T J S = S J S^T = J$$

where J is the "standard symplectic matrix" defined by

$$J = \begin{pmatrix} 0_{n\times n} & I_{n\times n} \\ -I_{n\times n} & 0_{n\times n} \end{pmatrix}.$$

Since

$$\det S^T J S = \det S^2 \det J = \det J,$$

it follows that $\det S$ can, a priori, take any of the two values ± 1. It turns out, however, that

$$S \in \mathrm{Sp}(n) \Longrightarrow \det S = 1.$$

Here is an algebraic proof of this important property. Recall that, to every antisymmetric matrix A, one associates a polynomial $\mathrm{Pf}(A)$ in the entries of A ("the Pfaffian of A"). The Pfaffian has the following properties:

$$\mathrm{Pf}(S^T A S) = (\det S)\,\mathrm{Pf}(A), \quad \mathrm{Pf}(J) = 1.$$

Choose now $A = J$ and $S \in \mathrm{Sp}(n)$. Since $S^T J S = J$, we have

$$\mathrm{Pf}(S^T J S) = \det S = 1,$$

which was to be proven.

1.5.3 Symplectic block-matrices

It is often helpful to write symplectic matrices in the block form

$$S = \begin{pmatrix} A & B \\ C & D \end{pmatrix} \tag{1.13}$$

where A, B, C, D are $n \times n$ matrices. The condition $S \in \mathrm{Sp}(n)$ is then equivalent to the two following sets of equivalent conditions:

$$A^T C,\ B^T D \text{ symmetric, and } A^T D - C^T B = I \tag{1.14}$$

$$AB^T,\ CD^T \text{ symmetric, and } AD^T - BC^T = I, \tag{1.15}$$

and it follows from the second of these sets of conditions that the inverse of S is

$$S^{-1} = \begin{pmatrix} D^T & -B^T \\ -C^T & A^T \end{pmatrix}. \tag{1.16}$$

Note that the monomorphism (1.12) is an isomorphism of the unitary group $\mathrm{U}(n, \mathbb{C})$ on the subgroup $\mathrm{U}(n)$ of $\mathrm{Sp}(n)$ consisting of all symplectic matrices

$$U = \begin{pmatrix} A & -B \\ B & A \end{pmatrix}. \tag{1.17}$$

Proposition 10. *The group monomorphism* $\iota : GL(n, \mathbb{C}) \longrightarrow GL(2n, \mathbb{R})$ *defined by*

$$\iota(A + iB) = \begin{pmatrix} A & -B \\ B & B \end{pmatrix} \tag{1.18}$$

restricts to a monomorphism $U(n, \mathbb{C}) \longrightarrow \mathrm{Sp}(n)$ *such that* $\iota(u^*) = \iota(u)^T$. *That monomorphism identifies the unitary group* $U(n, \mathbb{C})$ *with the subgroup* $U(n) = \iota(U(n, \mathbb{C}))$ *of* $\mathrm{Sp}(n)$.

Proof. It follows from conditions (1.14), (1.15) that

$$\iota(u) = U = \begin{pmatrix} A & -B \\ B & A \end{pmatrix} \tag{1.19}$$

is in $\mathrm{Sp}(n)$ if and only if

$$AB^T = B^T A, \quad AA^T + BB^T = I, \tag{1.20}$$

or, equivalently

$$A^T B = BA^T, \quad A^T A + B^T B = I. \tag{1.21}$$

The equivalence of conditions (1.20) and (1.21) is proved by noting that $U \in U(n)$ if and only if $U^T \in U(n)$ which follows from the fact that the monomorphism ι satisfies $\iota(u^*) = \iota(u)^T$. □

1.5.4 The eigenvalues of a symplectic matrix

The following result is well-known:

Proposition 11. *Let $S \in \mathrm{Sp}(n)$.* (i) *If λ is an eigenvalue of S, then so are $\bar{\lambda}$ and $1/\lambda$ (and hence also $1/\bar{\lambda}$);* (ii) *If the eigenvalue λ of S has multiplicity k, then so has $1/\lambda$.* (iii) *S and S^{-1} have the same eigenvalues.*

Proof. Let us show that the characteristic polynomial $P_S(\lambda) = \det(S - \lambda I)$ of S satisfies the reflexivity relationship

$$P_S(\lambda) = \lambda^{2n} P_S(1/\lambda). \tag{1.22}$$

Property (i) will follow since, for real matrices, eigenvalues appear in conjugate pairs. Since $S^T JS = J$, we have $S = -J(S^T)^{-1}J$, and hence

$$\begin{aligned} P_S(\lambda) &= \det(-J(S^T)^{-1}J - \lambda I) \\ &= \det(-(S^T)^{-1}J + \lambda I) \\ &= \det(-J + \lambda S) \\ &= \lambda^{2n} \det(S - \lambda^{-1} I) \end{aligned}$$

which is precisely (1.22). (ii). Let $P_S^{(j)}$ be the j-th derivative of the polynomial P_S. If λ_0 has multiplicity k, then $P_S^{(j)}(\lambda_0) = 0$ for $0 \le j \le k-1$ and $P_S^{(k)}(\lambda) \neq 0$. In view of (1.22), we also have $P_S^{(j)}(1/\lambda) = 0$ for $0 \le j \le k-1$ and $P_S^{(k)}(1/\lambda) \neq 0$. Property (iii) immediately follows from (ii). □

1.5.5 Williamson's symplectic diagonalization

Every real symmetric matrix can be diagonalized using an orthogonal transformation. When M is $2n{\times}2n$ positive definite symmetric matrix, then M can be diagonalized using a symplectic matrix. This is Williamson's famous symplectic diagonalization theorem. It has many applications throughout quantum mechanics and quantum information theory. We first remark that the eigenvalues of JM are the same as those of the antisymmetric matrix $M^{1/2}JM^{1/2}$; they are thus of the type $\pm i\lambda_j$, where $\lambda_j > 0$. The set of positive numbers $\{\lambda_j : 1 \le j \le n\}$ is called the *symplectic spectrum* of the matrix M. The symplectic spectrum is generally not related to the set of eigenvalues of M in any simple way.

Proposition 12. *Let M be a real symmetric positive definite $2n \times 2n$ matrix. There exists $S \in \mathrm{Sp}(n)$ such that $M = S^T DS$, where D is a diagonal matrix of the type*

$$D = \begin{pmatrix} \Lambda & 0 \\ 0 & \Lambda \end{pmatrix}$$

the diagonal entries of Λ *being the positive numbers* λ_j *such that* $\pm i\lambda_j$ *is an eigenvalue of* JM.

Proof. Let $\langle\cdot,\cdot\rangle_M$ be the scalar product on $\mathbb{C}^{2n}$ defined by $\langle z,z'\rangle_M = \langle Mz,z'\rangle$. Since both $\langle\cdot,\cdot\rangle_M$ and the symplectic form σ are non-degenerate, we can find a unique invertible matrix K of order $2n$ such that

$$\langle z,Kz'\rangle_M = \sigma(z,z')$$

for all z,z'; that matrix satisfies $K^TM = J = -MK$. Since σ is antisymmetric, we must have $K = -K^M$, where $K^M = -M^{-1}K^TM$ is the transpose of K with respect to $\langle\cdot,\cdot\rangle_M$; it follows that the eigenvalues of $K = -M^{-1}J$ are of the type $\pm i\lambda_j$, $\lambda_j > 0$, and so are those of JM^{-1}. The corresponding complex eigenvectors occurring in conjugate pairs $e'_j \pm if'_j$ we thus obtain a $\langle\cdot,\cdot\rangle_M$-orthonormal basis $\{e'_i,f'_j\}_{1\leq i,j\leq n}$ of $\mathbb{R}^{2n}$ such that $Ke'_i = \lambda_i f'_i$ and $Kf'_j = -\lambda_j e'_j$. It follows from these relationships that we have $K^2e'_i = -\lambda_i^2 e'_i$ and $K^2f'_j = -\lambda_j^2 f'_j$ and that the vectors of the basis $\{e'_i,f'_j\}_{1\leq i,j\leq n}$ satisfy the relationships

$$\begin{aligned}
\sigma(e'_i,e'_j) &= \langle e'_i,Ke'_j\rangle_M = \lambda_j\langle e'_i,f'_j\rangle_M = 0\\
\sigma(f'_i,f'_j) &= \langle f'_i,Kf'_j\rangle_M = -\lambda_j\langle f'_i,e'_j\rangle_M = 0\\
\sigma(f'_i,e'_j) &= \langle f'_i,Ke'_j\rangle_M = \lambda_i\langle f'_i,f'_j\rangle_M = -\lambda_i\delta_{ij}.
\end{aligned}$$

Setting $e_i = \lambda_i^{-1/2}e'_i$ and $f_j = \lambda_j^{-1/2}f'_j$, the basis $\{e_i,f_j\}_{1\leq i,j\leq n}$ is symplectic. Let S be the element of $\mathrm{Sp}(n)$ mapping the canonical symplectic basis to $\{e_i,f_j\}_{1\leq i,j\leq n}$. The $\langle\cdot,\cdot\rangle_M$-orthogonality of $\{e_i,f_j\}_{1\leq i,j\leq n}$ implies $M = S^TDS$ with $D = \mathrm{diag}(\Lambda,\Lambda)$, $\Lambda = \mathrm{diag}(\lambda_1,\dots,\lambda_n)$. □

1.6 Quantum mechanics

For George Mackey, quantum mechanics is a refinement of classical mechanics in its Hamiltonian form. It is thus viewed as a “super theory” containing classical mechanics if certain limiting conditions are satisfied (for instance when Planck’s constant can be neglected). This is the point of view we have adopted in our book *Emergence of the Quantum from the Classical: Mathematical Aspects of Quantum Processes* (World Scientific, 2017), where the point of view is also maintained that there is a form of emergence of the quantum from the classical. Quantum mechanics has a double status. While it has its roots in physics, it has become (and this rather rapidly since its inception) a branch of both pure and applied mathematics. That applied mathematics should play a role is rather obvious since there are many practical and numerical issues to be solved, but parts of the problems that are posed by quantum mechanics definitely belong to more abstract areas (this is also true, of course, of Hamiltonian mechanics, which has led to important developments in symplectic topology and geometry; Gromov’s non-squeezing theorem is a well-known example). Anyway, it is clear that mathematics has largely contributed to the growth and understanding of quantum mechanics, and this from the very beginning.

1.6.1 The axioms of quantum mechanics

In the title of this section, we use the term “axiom” instead of “postulate”. This is because we want to carefully avoid the quagmire of defining the rules governing quantum mechanics as a *physical theory;* we leave this ungrateful task to physicists and philosopher's of science. The debate is open, and is not, for sure, to be soon closed. The first six axioms are rather standard and are essentially those of the “Copenhagen School” (they are actually due to John von Neumann). For a nice discussion, see for instance Paul Thierry [61]:

Axiom 13. *A pure quantum state of a physical system S is an element $\psi \neq 0$ of a separable complex Hilbert space H_S, called the state space of S. Two normalized elements ψ and ϕ of H_S define the same state if there exists $c \in \mathbb{C}$ $(c \neq 0)$ such that $\psi = c\phi$.*

Equivalently, a quantum state is a ray (= one dimensional subspace) $\mathbb{C}\psi$ in H_S. A finite sum of states $\psi_1 + \cdots + \psi_m$, with $\psi_j \in H_S$, $1 \leq j \leq m$, is also a state called the superposition of the states $\psi_1, \ldots, \psi_m$.

Axiom 14. *A (mixed) quantum state in S is the datum of a set $\{(\psi_j, \alpha_j) \in H_S \times \mathbb{R}_+ : j \in F\}$ with $\|\psi_j\|_H = 1$ and where F is a countable set and $\sum_{j\in F} \alpha_j = 1$; α_j is the probability that the quantum state is the pure state ψ_j.*

Axiom 15. *An observable of a quantum system S with state space H_S is a (possibly unbounded) self-adjoint operator $\widehat{A}$ on H_S.*

This axiom has to be complemented by the following statistical hypothesis, called “Born's rule” (the validity of which is being regularly questioned by physicists):

Axiom 16. *Born's rule: if $\lambda_j \in \mathbb{R}$ and a normalized $\psi_j \in H_S$ are such that $\widehat{A}\psi_j = \lambda_j\psi_j$, then, if the system S is in the state ψ, the observable $\widehat{A}$ has the probability $|(\psi|\psi_j)_H|^2$ of taking the value λ_j.*

So far, the listed axioms define “standard” quantum mechanics. We will complete these axioms to enlarge the picture to “quantum mechanics in phase space”. For this purpose, we assume that $H_S = L^2(\mathbb{R}^n)$.

Axiom 17. *The datum of a mixed quantum state $\{(\psi_j, \alpha_j) \in L^2(\mathbb{R}^n) \times \mathbb{R}_+ : j \in F\}$ is equivalent to that of its Wigner distribution $\rho = \sum_{j\in F} \alpha_j W\psi_j$.*

Axiom 18. *Suppose the quantum system S is in the mixed state $\{(\psi_j, \alpha_j) \in L^2(\mathbb{R}^n) \times \mathbb{R}_+ : j \in F\}$, and let $\widehat{A}$ be a quantum observable. The expectation (or: average) value of $\widehat{A}$ in this mixed state is given, when defined, by the integral $\int_{\mathbb{R}^{2n}} a(z)\rho(z)dz$.*

1.6.2 Quantization

In this book, we will not really use the quantum mechanical axioms just listed; we will actually be focusing on another related issue, that of the "quantization of observables". Mathematically speaking, a quantization is a procedure $a \longmapsto \mathrm{Op}(a)$ that associates to a symbol $a \in \mathcal{S}'(\mathbb{R}^{2n})$ an operator $\widehat{A} = \mathrm{Op}(a)$ (in general a pseudodifferential operator $\mathcal{S}(\mathbb{R}^n) \longrightarrow \mathcal{S}'(\mathbb{R}^n)$). This procedure should satisfy the following requirements to qualify as a quantization:

1. **Continuity**: Op is a continuous linear mapping
$$\mathcal{S}'(\mathbb{R}^{2n}) \longrightarrow \mathcal{L}(\mathcal{S}(\mathbb{R}^n), \mathcal{S}'(\mathbb{R}^n))\,;$$
2. **Triviality**: Op should satisfy
$$\mathrm{Op}(b \otimes 1)u = bu, \quad \mathrm{Op}(1 \otimes f)u = \mathcal{F}^{-1}(f\mathcal{F}u) \tag{1.23}$$
for $\psi \in \mathcal{S}(\mathbb{R}^n)$ and $b \in \mathcal{S}'(\mathbb{R}^n)$,
3. **Self-adjointness**:
$$\mathrm{Op}(a) \text{ is self-adjoint} \Longleftrightarrow a \text{ is real}; \tag{1.24}$$
4. **"Dirac's dream"**:
$$[\mathrm{Op}(a), \mathrm{Op}(b)] = i\hbar\, \mathrm{Op}(\{a, b\}) \tag{1.25}$$
where $[\cdot,\cdot]$ is the commutator and $\{a, b\}$ the Poisson bracket associated with the standard symplectic form, i. e., $\{a, b\} = \partial_x a \cdot \partial_p b - \partial_x b \cdot \partial_p a$.

While the first three conditions are satisfied by all known useful quantizations, the fourth property, "Dirac's dream", is not. Actually, the impossibility of constructing a full quantization satisfying all for properties is the famous result of Groenewold and van Hove, which is a "no-go" result. Mathematically, it says (in its strong form) that one cannot quantize the Poisson algebra of polynomials in $\mathbb{R}^n$, beyond those of degree ≤ 2.

1.6.3 Superposition and entanglement

The typically "quantum" notions of superposition and entanglement are both surrounded by an almost mystical aura in nonscientific circles. However, both notions are perfectly well-defined mathematically. Here is a short and very elementary description, mainly addressed to mathematicians who might not be familiar with these concepts.

A pure classical state is a point in some phase space, and observables are functions defined on this phase space. In quantum mechanics, a pure state is a point in a

Hilbert space, and observables are self-adjoint operators[1] acting on that Hilbert space. In quantum physics, these observables are called "Hermitian operators"; it should be noted that in the physics literature the distinction between operators that are merely symmetric and operators that are actually self-adjoint is generally glossed over. Why does one insist that pure states should belong to a Hilbert space? It is because in Hilbert spaces we can define unambiguously orthogonal projections on subspaces (or more generally, on closed convex subsets); it is the completeness property of a Hilbert space that allows us to construct these projections by approximating them with Cauchy sequences. This property is essential when one wants to discuss the collapse (or: reduction) of the state function. Suppose indeed that we are studying a quantum observable $\widehat{A}$ on the Hilbert space $L^2(\mathbb{R}^n)$; we denote the scalar product of two elements ψ, ϕ of $L^2(\mathbb{R}^n)$ by $(\psi|\phi)_{L^2}$. We assume $\widehat{A}$ has a discrete spectrum $\lambda_1, \lambda_2, \ldots$ with corresponding orthonormal eigenfunctions $\psi_1, \psi_2, \ldots$ of $L^2(\mathbb{R}^n)$. If measurements are performed on the state, then the only values that will ever be observed are the eigenvalues λ_j. However, this does not mean that the observable was initially in the pure state ψ_j; the most general pure state corresponding to that observable is namely a superposition $\psi = \sum_j (\psi|\psi_j)_{L^2}\psi_j$. In view of Plancherel's formula, we have $\sum_j |(\psi|\psi_j)_{L^2}|^2 = 1$ which allows us to view the numbers $|(\psi|\psi_j)_{L^2}|^2$ as probabilities: According to all standard interpretations of quantum mechanics, $|(\psi|\psi_j)_{L^2}|^2$ is the probability that the measurement performed on the state will yield the value λ_j. Now comes the subtle point: After this measurement, the new state will be a linear combination of the eigenfunctions $\psi_j^{(1)}, \ldots, \psi_j^{(m)}$ corresponding to λ_j (m the multiplicity of λ_j); the measurement thus orthogonally projects ("collapses") the unknown state ψ onto the subspace spanned by the set of orthonormal vectors $\{\psi_j^{(1)}, \ldots, \psi_j^{(m)}\}$.

1.6.4 Mixed quantum states; the density matrix

In practice, it is usual not possible to assign a pure state to the system being studied because of uncertainties of a statistical nature. It is here that the notion of mixed state intervenes. In classical mechanics, a mixed state is just a probability distribution ("density of states"), i. e., a function $\rho \geq 0$ on phase space whose integral is one with respect to some measure defined on that phase space. This function may be compactly supported, or not. In quantum mechanics, the situation is not so much different. A mixed quantum state is represented by a quasi-probability distribution ρ defined on phase space (the "Wigner distribution"), the prefix "quasi" meaning that ρ can (and usually does) take negative values. For all practical purposes, this function

1 In quantum mechanics, the observables are called "Hermitian operators"; it should be noted that in the physics literature the distinction between operators that are merely symmetric and operators that are actually self-adjoint is generally ignored.

ρ is used to calculate statistical averages exactly as if it were a *bona fide* probability distribution. Suppose indeed the quantum mixed state consists of a collection of pairs $\{(\psi_j, \alpha_j)\}$ where the $\psi_j \in L^2(\mathbb{R}^n)$ are pure states, and the α_j are positive numbers summing up to one. Physically this corresponds to representing the degree of knowledge of the state: the probability that it is in the pure state ψ_j is α_j. A mixed state is thus a statistical mixture; it is different from a linear combination as just considered. A mixed state $\{(\psi_j, \alpha_j)\}$ can be represented advantageously in two different but mathematically equivalent ways. The first is to identify $\{(\psi_j, \alpha_j)\}$ with its Wigner distribution ρ: by definition

$$\rho(x,p) = \sum \alpha_j W\psi_j(x,p)$$

where $W\psi_j$ is the Wigner transform of ψ_j:

$$W\psi_j(x,p) = \left(\frac{1}{2\pi\hbar}\right)^n \int_{\mathbb{R}^n} e^{-\frac{i}{\hbar}py} \psi_j\left(x + \frac{1}{2}y\right) \overline{\psi_j\left(x - \frac{1}{2}y\right)} dy. \tag{1.26}$$

The second way to represent $\{(\psi_j, \alpha_j)\}$ is to note that, since the datum of a pure state ψ_j is equivalent to the datum of the orthogonal projection $\widehat{\Pi}_j\psi = (\psi|\psi_j)_{L^2}\psi_j$, we can identify the mixed state with the convex sum of projectors $\widehat{\rho} = \sum_j \alpha_j \widehat{\Pi}_j$. The operator $\widehat{\rho}$ is called the *density operator (or matrix)* of the mixed state $\{(\psi_j, \alpha_j)\}$. It is a bounded, positive, and self-adjoint operator on $L^2(\mathbb{R}^n)$; it moreover has trace $\operatorname{Tr}(\widehat{\rho}) = 1$. It turns out that both representations are equivalent because $\widehat{\rho}$ is just the Weyl operator whose symbol is $(2\pi\hbar)^n\rho$, that is

$$\widehat{\rho}\psi(x) = \int_{\mathbb{R}^{2n}} e^{\frac{i}{\hbar}p(x-y)} \rho\left(\frac{1}{2}(x+y), p\right) \psi(y) dy dp.$$

1.6.5 Entanglement: spooky actions at a distance

Assume that we are dealing with two quantum systems, labelled A and B, whose Hilbert spaces are H^A and H^B. To each of these systems, we assign an “observer”: Alice for A, and Bob for B physicists like Alice and Bob; if there is a third observer, he is called Charles. The tasks of these observers will be to perform measurements, either on H^B or on H^B, or on both. Suppose first that $\dim H^A = \dim H^B = 1$; then the state of A is a pure state ψ^A and, similarly, the state of B is a pure state ψ^B. We now consider the union of the two systems A and B; it is a new quantum system $A \otimes B$ with one-dimensional Hilbert space $H^A \otimes H^B$ consisting of all functions $c_{AB}(\psi^A \otimes \psi^B)$ with $c_{AB} \in \mathbb{C}$. If now Alice observes the part A of the state $A \otimes B$, she will collapse $\psi^A \otimes \psi^B$ to ψ^A, and Bob will, similarly, collapse $\psi^A \otimes \psi^B$ to ψ^B. Nothing surprising here, of course. But suppose now $\dim H^A = \dim H^B = 2$, and choose orthonormal bases (ψ^A, ϕ^A) of

H^A and (ψ^B, ϕ^B) of H^B. The system A is thus in some normalized state $a_A\psi^A + c_A\phi^A$, and the system B in some normalized state $a_B\psi^B + c_B\phi^B$. Consider again the global system $A \otimes B$; its Hilbert space $H^A \otimes H^B$ is four dimensional, and has an orthonormal basis $(\psi^A \otimes \psi^B, \psi^A \otimes \phi^B, \phi^A \otimes \psi^B, \phi^A \otimes \phi^B)$, so the most general state of $A \otimes B$ is here an arbitrary normalized linear combination of these vectors. Consider in particular the state

$$\psi = \frac{1}{\sqrt{2}}(\psi^A \otimes \phi^B + \phi^A \otimes \psi^B) \tag{1.27}$$

(such a state is called a "Bell state" in the physical literature). We now let Alice perform a measurement on the subsystem A of $A \otimes B$. There are two possible outcomes, each having the same probability 50 %: Either (i) Alice measures ψ^A and hence ψ collapses to $\psi^A \otimes \phi^B$, and in any subsequent measurement Bob will find B in the state ϕ^B or (ii) she measures ϕ^A, in which case ψ collapses to $\phi^A \otimes \psi^B$, and Bob will automatically measure ψ^B. The point here is that Alice's measurement is random, but her act of observation determines the outcome Bob will observe. This very non-classical aspect of quantum mechanics was considered to be shocking, led to the famous EPR paper [24] by Einstein, Podolsky, and Rosen, and also led Einstein to speak about "spooky actions at a distance".

1.6.6 Variable $\hbar$

In quantum physics, it does not make sense to speak about the exact value of h; it is a constant of nature, not a mathematical constant like π, e, or $\sqrt{2}$. Its value not only depends on the choice of units, but also on measurements. Fixing once and for all a system of units, all we can say is that h is an experimental value situated in some real interval $[\hbar_0 - \varepsilon, \hbar_0 - \varepsilon]$ (this is equally true of any other physical constant). For instance, on 20 May 2019, the SI unit of mass (the kilogram) was redefined by the BIPM by fixing *arbitrarily* the value of Planck's constant as being

$$h = 6.626070150 \times 10^{-34}\,\mathrm{J\,s}. \tag{1.28}$$

This choice is, however, *ad hoc*, and was meant to make the kilogram fit with its best known values. However, measurements given by the National Institute of Standards (NIST) in 2014 yield the interval of confidence

$$h = 6.626070150(81) \times 10^{-34}\,\mathrm{J\,s}; \tag{1.29}$$

more recent measurements, also performed at NIST, yield the *different* result

$$h = 6.626070040(81) \times 10^{-34}\,\mathrm{J\,s}. \tag{1.30}$$

A few historical remarks. Paul Dirac [21] suggested in 1937 in his "Large Numbers Hypothesis" that some constants of nature could vary in space-time. Since then, the topic has remained a subject of fascination which has motivated numerous theoretical and experimental researches. The variability of physical parameters (we are avoiding the oxymoron "nonconstant constants") is a possibility that cannot be outruled and which has being actively studied for a long time by many physicists. Dirac had speculated that physical constants, such as the gravitational constant or the fine-structure constant, might be subject to change over time. However, testing the constancy of a physical parameter means going to extraordinary lengths in terms of precision measurements and is intimately related to choices of *unit systems*. The history actually started in a quite romantic way, with the story of the Oklo natural nuclear reactor found in a uranium mine in Central Africa in 1972. The measurements that were made in Oklo give limits on the variation of the fine-structure constant over the period since the reactor was running for *ca*. 1.8 billion years, which is much less than the estimated age of universe. In 1999, a team of astronomers headed by the astrophysicist John Webb reported that measurements of light absorbed by very distant quasars suggest that the value of the fine-structure constant was once slightly different from what it is today. These experiments, made using the very sophisticated Keck and VLT telescopes in Hawaii, put an upper bound on the relative change per year, at roughly 10^{-17} per year.

2 Displacements and reflections

In this chapter, we introduce the most basic tools from quantum harmonic analysis: the Heisenberg displacement operator and the Grossmann–Royer reflection operator. They are the building blocks of the operator theories we will study in the forthcoming chapters.

2.1 The Heisenberg displacement operator

2.1.1 Definition and motivation

In what follows, ψ is a square integrable complex function: $\psi \in L^2(\mathbb{R}^n)$. We recall that the Schwartz space $\mathcal{S}(\mathbb{R}^n)$ is dense in $L^2(\mathbb{R}^n)$, so in many proofs it is sufficient to assume $\psi \in \mathcal{S}(\mathbb{R}^n)$, and then to conclude using a continuity and density argument.

Definition 19. Let $z_0 = (x_0, p_0)$ be in $\mathbb{R}^{2n} \equiv \mathbb{R}^n \times \mathbb{R}^n$. The unitary operator

$$\widehat{D}(z_0) : L^2(\mathbb{R}^n) \longrightarrow L^2(\mathbb{R}^n)$$

defined by

$$\widehat{D}(z_0)\psi(x) = e^{\frac{i}{\hbar}(p_0\cdot x - \frac{1}{2}p_0\cdot x_0)}\psi(x - x_0) \tag{2.1}$$

is called the (Heisenberg) displacement operator determined by z_0.

Another common denomination for $\widehat{D}(z_0)$ is “Heisenberg–Weyl operator”; physicists sometimes even use the shortened terminology “Weyl operator”, but this is confusing because the notion of Weyl operator has a very different meaning in mathematics.

Admittedly, this definition seems to be *ad hoc* and somewhat mysterious. Here is a dynamical motivation. Let us begin with some rather trite considerations. The phase space translation operators

$$T(z_0) : z \longmapsto z + z_0$$

can be viewed as the time-one map of a Hamiltonian flow, namely that determined by the Hamiltonian function

$$H(z) = \sigma(z, z_0) = p \cdot x_0 - p_0 \cdot x. \tag{2.2}$$

In fact, the solutions of the associated Hamilton equations $\dot{z} = J\partial_z H(z)$ are given by $z(t) = z + tz_0$. Consider now the “quantized” version of H; it is the operator

$$\widehat{H} = x_0 \cdot (-i\hbar\partial_x) - p_0 \cdot x.$$

https://doi.org/10.1515/9783110722772-002

One can check by a straightforward calculation that the solution ψ of the associated Schrödinger equation

$$i\hbar\partial_t\psi = \widehat{H}\psi, \quad \psi(\cdot,0) = \psi_0$$

is given by the formula

$$\psi(x,t) = e^{\frac{i}{\hbar}(tp_0\cdot x - \frac{1}{2}t^2 p_0\cdot x_0)}\psi_0(x - tx_0), \tag{2.3}$$

that is

$$\psi(\cdot,t) = \widehat{D}(tz_0)\psi_0.$$

The Heisenberg displacement operator is thus the time-one propagator for the Schrödinger equation associated with the displacement Hamiltonian. This justifies the notation

$$\widehat{D}(z_0) = e^{\frac{i}{\hbar}\sigma(z_0,\hat{z})}$$

that is often used in the literature.

2.1.2 Properties

The displacement operators are linear:

$$\widehat{D}(z_0)(\lambda\psi + \mu\phi) = \lambda\widehat{D}(z_0)\psi + \mu\widehat{D}(z_0)\phi$$

and unitary: the inverse of $\widehat{D}(z_0)$ is

$$\widehat{D}(z_0)^{-1} = \widehat{D}(-z_0) = \widehat{D}(z_0)^*. \tag{2.4}$$

In fact, setting $x' = x - x_0$,

$$\begin{aligned}(\widehat{D}(z_0)\psi|\phi) &= \int e^{\frac{i}{\hbar}(p_0\cdot x - \frac{1}{2}p_0\cdot x_0)}\psi(x - x_0)\overline{\phi(x)}d^n x \\ &= \int \psi(x - x_0)\overline{e^{-\frac{i}{\hbar}(p_0\cdot x - \frac{1}{2}p_0\cdot x_0)}\phi(x)}d^n x \\ &= \int \psi(x')\overline{e^{\frac{i}{\hbar}(-p_0\cdot x - \frac{1}{2}p_0\cdot x_0)}\phi(x' + x_0)}d^n x \\ &= (\psi|\widehat{D}(-z_0)\phi).\end{aligned}$$

We leave the proof of following property to the reader.

Proposition 20. *The restriction of $\widehat{D}(z_0)$ to the Schwartz space $\mathcal{S}(\mathbb{R}^n)$ is a continuous automorphism $\mathcal{S}(\mathbb{R}^n) \longrightarrow \mathcal{S}(\mathbb{R}^n)$ that extends by duality into a continuous automorphism of $\mathcal{S}'(\mathbb{R}^n)$.*

The displacement operators do not commute; in fact, they satisfy the following remarkable properties

Proposition 21. *The displacement operators satisfy*

$$\widehat{D}(z_0)\widehat{D}(z_1) = e^{\frac{i}{\hbar}\sigma(z_0,z_1)}\widehat{D}(z_1)\widehat{D}(z_0) \tag{2.5}$$

$$\widehat{D}(z_0 + z_1) = e^{-\frac{i}{2\hbar}\sigma(z_0,z_1)}\widehat{D}(z_0)\widehat{D}(z_1) \tag{2.6}$$

for all $z_0, z_1 \in \mathbb{R}^{2n}$.

Proof. We have

$$\begin{aligned}\widehat{D}(z_0)\widehat{D}(z_1) &= \widehat{D}(z_0)(e^{\frac{i}{\hbar}(p_1\cdot x - \frac{1}{2}p_1\cdot x_1)}T(z_1)) \\ &= e^{\frac{i}{\hbar}(p_0\cdot x - \frac{1}{2}p_0\cdot x_0)}e^{\frac{i}{\hbar}(p_1\cdot(x-x_0) - \frac{1}{2}p_1\cdot x_1)}T(z_0 + z_1)\end{aligned}$$

and, similarly

$$\widehat{D}(z_1)\widehat{D}(z_0) = e^{\frac{i}{\hbar}(p_1\cdot x - \frac{1}{2}p_1\cdot x_1)}e^{\frac{i}{\hbar}(p_0\cdot(x-x_1) - \frac{1}{2}p_0\cdot x_0)}T(z_0 + z_1).$$

Setting

$$\begin{aligned}\Phi &= p_0 \cdot x - \frac{1}{2}p_0 \cdot x_0 + p_1 \cdot (x - x_0) - \frac{1}{2}p_1 \cdot x_1 \\ \Phi' &= p_1 \cdot x - \frac{1}{2}p_1 \cdot x_1 + p_0 \cdot (x - x_1) - \frac{1}{2}p_0 \cdot x_0;\end{aligned}$$

we have

$$\widehat{D}(z_0)\widehat{D}(z_1) = e^{\frac{i}{\hbar}(\Phi-\Phi')}\widehat{D}(z_1)\widehat{D}(z_0)$$

which yields

$$\Phi - \Phi' = p_0 \cdot x_1 - p_1 \cdot x_0 = \sigma(z_0, z_1)$$

which proves (2.5). Let us next prove formula (2.6). We have

$$\widehat{D}(z_0 + z_1) = e^{\frac{i}{\hbar}\Phi''}T(z_0 + z_1)$$

with

$$\Phi'' = (p_0 + p_1) \cdot x - \frac{1}{2}(p_0 + p_1) \cdot (x_0 + x_1).$$

On the other hand, we have seen previously that

$$\widehat{D}(z_0)\widehat{D}(z_1) = e^{\frac{i}{\hbar}\Phi}T(z_0 + z_1),$$

so that

$$\widehat{D}(z_0 + z_1) = e^{\frac{i}{\hbar}(\Phi'' - \Phi)}\widehat{D}(z_0)\widehat{D}(z_1).$$

A straightforward algebraic calculation shows that

$$\Phi'' - \Phi = \frac{1}{2}p_1 \cdot x_0 - \frac{1}{2}p_0 \cdot x_1 = -\frac{1}{2}\sigma(z_0, z_1)$$

hence formula (2.6). □

Proposition 22. (i) *The mapping $z_0 \longmapsto \widehat{D}(z_0)$ is strongly continuous on $\mathcal{S}(\mathbb{R}^n)$, that is*

$$\lim_{z \to z_0} \|\widehat{D}(z)\psi - \widehat{D}(z_0)\psi\|'_{\alpha,\beta} = 0$$

for every $\psi \in \mathcal{S}(\mathbb{R}^n)$, and $\alpha, \beta \in \mathbb{N}^n$ where $\|\cdot\|'_{\alpha,\beta}$ $((\alpha,\beta) \in \mathbb{N}^n \times \mathbb{N}^n)$ is the seminorm

$$\|\psi\|'_{\alpha,\beta} = \sup|\partial_x^\alpha x^\beta \psi|$$

on $\mathcal{S}(\mathbb{R}^n)$; (ii) *The mapping $z_0 \longmapsto \widehat{D}(z_0)$ is weakly $*$-continuous on $\mathcal{S}'(\mathbb{R}^n)$, that is*

$$\lim_{z \to z_0} \langle \widehat{D}(z)\psi, \phi \rangle = \langle \widehat{D}(z_0)\psi, \phi \rangle$$

for all $\psi \in \mathcal{S}'(\mathbb{R}^n)$ and $\phi \in \mathcal{S}(\mathbb{R}^n)$.

Proof. (i) Let us assume that $z_0 = 0$ and prove that

$$\lim_{|z| \to 0} \|\partial_x^\alpha x^\beta (\widehat{D}(z)\psi - \psi)\|_\infty = 0.$$

Noting that, by Leibniz' formula for the derivatives of a product, we can write

$$\partial_x^\alpha x^\beta \widehat{D}(z)\psi = \sum_{\gamma \le \alpha, \delta \le \beta} c_{\alpha\beta\gamma\delta} x^\delta p^\gamma \widehat{D}(z)(\partial_x^{\alpha-\gamma} x^{\beta-\delta}\psi) \tag{2.7}$$

where the $c_{\alpha\beta\gamma\delta}$ are complex constants and $\gamma \le \alpha$ means $\gamma_j \le \alpha_j$ for $j = 1, 2, \ldots, n$, we have

$$\|\partial_x^\alpha x^\beta (\widehat{D}(z)\psi - \psi)\|_\infty \le \|\widehat{D}(z)(\partial_x^\alpha x^\beta \psi) - (\partial_x^\alpha x^\beta \psi)\|_\infty + \sum_{\substack{0<\gamma\le\alpha \\ 0<\delta\le\beta}} c_{\alpha\beta\gamma\delta} |x^\delta p^\gamma| \|\widehat{D}(z)(\partial_x^{\alpha-\gamma} x^{\beta-\delta}\psi)\|_\infty.$$

Now, since $\gamma \ne 0$, $\delta \ne 0$, it is clear that

$$\lim_{z \to 0} |x^\delta p^\gamma| \, \|\widehat{D}(z)(\partial_x^{\alpha-\gamma} x^{\beta-\delta}\psi)\|_\infty = 0,$$

so there remains to show that

$$\lim_{z\to 0}\|\widehat{D}(z)(\partial_x^\alpha x^\beta\psi) - (\partial_x^\alpha x^\beta\psi)\|_\infty = 0.$$

That this holds if ψ is compactly supported, that is, $\psi \in C_0^\infty(\mathbb{R}^n)$, is clear. The equality for ψ in the general case then follows using the density of $C_0^\infty(\mathbb{R}^n)$ in $\mathcal{S}(\mathbb{R}^n)$. The case $z_0 \neq 0$ is treated similarly. (ii) It suffices again to assume $z_0 = 0$. Let $\psi \in \mathcal{S}'(\mathbb{R}^n)$ and $\phi \in \mathcal{S}(\mathbb{R}^n)$; we have

$$\langle \widehat{D}(x,p)\psi, \phi\rangle = \langle \psi, \widehat{D}(-x,p)\phi\rangle$$

and hence

$$\lim_{z\to 0}\langle \widehat{D}(x,p)\psi, \phi\rangle = \lim_{z\to 0}\langle \psi, \widehat{D}(-x,p)\phi\rangle = \langle \psi, \phi\rangle,$$

so that $\widehat{D}(z)$ is weakly $*$-continuous on $\mathcal{S}'(\mathbb{R}^n)$ as claimed. The case $z_0 \neq 0$ is treated similarly. □

2.2 The Grossmann–Royer reflection operators

The Grossmann–Royer reflection operators are by definition the conjugate by a displacement operator of the ordinary reflection (or parity) operator

$$\widehat{R} : L^2(\mathbb{R}^n) \longrightarrow L^2(\mathbb{R}^n), \quad \widehat{R}\psi(x) = \psi(-x).$$

Definition 23. The (Grossmann–Royer) reflection operator $\widehat{R}(z_0)$ is the operator

$$\widehat{R}(z_0) : L^2(\mathbb{R}^n) \longrightarrow L^2(\mathbb{R}^n)$$

defined by

$$\widehat{R}(z_0) = \widehat{D}(z_0)\widehat{R}\widehat{D}(z_0)^{-1}. \tag{2.8}$$

The operators $\widehat{R}(z_0)$ are also called “parity operators”, especially in the physical literature.

The reflection operators are obviously linear:

$$\widehat{R}(z_0)(\lambda\psi + \mu\phi) = \lambda\widehat{R}(z_0)\psi + \mu\widehat{R}(z_0)\phi.$$

Using the explicit expression (2.1) of the displacement operator, we get an explicit expression for $\widehat{R}$: Let $\psi \in L^2(\mathbb{R}^n)$ (or $\mathcal{S}(\mathbb{R}^n)$ or $\mathcal{S}'(\mathbb{R}^n)$). A few calculations yield

$$\widehat{R}(z_0)\psi(x) = e^{\frac{2i}{\hbar}p_0\cdot(x-x_0)}\psi(2x_0 - x) \tag{2.9}$$

for $z_0 = (x_0, p_0)$. The verification of this identity is straightforward and left to the reader.

Since the Heisenberg displacement operators and the reflection $\widehat{R}$ are both unitary on $L^2(\mathbb{R}^n)$, the reflection $\widehat{R}(z_0)$ also are unitary (which also follows directly from formula (2.9)). The transpose $\widehat{R}(z_0)^T$ is also a reflection operator, and we have, for $z_0 = (x_0, p_0)$,

$$\widehat{R}(z_0)^T = \widehat{R}(\overline{z_0}), \quad \overline{z_0} = (x_0, -p_0). \tag{2.10}$$

Let us prove this equality. For $(\psi, \phi) \in \mathcal{S}(\mathbb{R}^n) \times \mathcal{S}(\mathbb{R}^n)$, we have, by definition,

$$\langle \widehat{R}(z_0)\psi, \phi\rangle = \langle \psi, \widehat{R}(z_0)^T \phi\rangle,$$

that is, setting $y = 2x_0 - x$,

$$\begin{aligned}\langle \widehat{R}(z_0)\psi, \phi\rangle &= \int_{\mathbb{R}^n} e^{\frac{2i}{\hbar}p_0\cdot(x-x_0)}\psi(2x_0 - x)\phi(x)dx \\ &= \int_{\mathbb{R}^n} e^{\frac{2i}{\hbar}p_0\cdot(x_0-x)}\psi(y)\phi(2x_0 - y)dy \\ &= \langle \psi, \widehat{R}(\overline{z_0})\phi\rangle.\end{aligned}$$

It follows from Proposition 20 that the restriction of $\widehat{R}(z_0)$ to the Schwartz space $\mathcal{S}(\mathbb{R}^n)$ is a continuous linear automorphism $\mathcal{S}(\mathbb{R}^n) \longrightarrow \mathcal{S}(\mathbb{R}^n)$ that extends by duality into a continuous automorphism of $\mathcal{S}'(\mathbb{R}^n)$.

The operator $\widehat{R}(z_0)$ is an involution that is

$$\widehat{R}(z_0)\widehat{R}(z_0) = I$$

where I is the identity operator on $L^2(\mathbb{R}^n)$ (or on $\mathcal{S}(\mathbb{R}^n)$ or $\mathcal{S}'(\mathbb{R}^n)$): since $\widehat{R}^2 = I$, we have

$$\widehat{R}(z_0)\widehat{R}(z_0) = \widehat{D}(z_0)\widehat{R}\widehat{D}(z_0)^{-1}\widehat{D}(z_0)\widehat{R}\widehat{D}(z_0)^{-1} = I.$$

The reflection operators satisfy the following product formula,

Proposition 24. *The operators $\widehat{R}(z_0)$ satisfy the product formula*

$$\widehat{R}(z_0)\widehat{R}(z_1) = e^{-\frac{2i}{\hbar}\sigma(z_0,z_1)}\widehat{T}(2(z_0 - z_1)) \tag{2.11}$$

for all $z_0, z_1 \in \mathbb{R}^{2n}$.

Proof. We have

$$\widehat{R}(z_0)\widehat{R}(z_1)\psi(x) = \widehat{R}(z_0)[e^{\frac{2i}{\hbar}p_1\cdot(2x_0-x-x_1)}\psi(2x_1 - x)]$$

$$= e^{\frac{2i}{\hbar}p_0\cdot(x-x_0)}e^{\frac{2i}{\hbar}p_1\cdot(x-x_1)}\psi(2x_1-(2x_0-x))$$
$$= e^{\frac{i}{\hbar}\Phi}\psi(x-2(x_0-x_1))$$

with

$$\Phi = 2[(p_0-p_1)x - p_0x_0 - p_1x_1 + 2p_1x_0].$$

On the other hand,

$$\widehat{T}(2(z_0-z_1))\psi(x) = e^{\frac{i}{\hbar}\Phi'}\psi(x-2(x_0-x_1))$$

with

$$\Phi' = 2((p_0-p_1)x - (p_0-p_1)(x_0-x_1)).$$

We have $\Phi - \Phi' = -2\sigma(z_0, z_1)$, hence the result. □

2.3 The symplectic Fourier transform

The symplectic Fourier transform is a variant of the usual Fourier transform on $\mathbb{R}^{2n}$.

Definition 25. The symplectic Fourier transform F_σ is defined, for $a \in \mathcal{S}(\mathbb{R}^{2n})$, by

$$a_\sigma(z) = F_\sigma a(z) = \left(\frac{1}{2\pi\hbar}\right)^n \int_{\mathbb{R}^{2n}} e^{-\frac{i}{\hbar}\sigma(z,z')}a(z')dz' \tag{2.12}$$

where $\sigma(z,z') = \sum_{j=1}^n p_jx_j' - p_j'x_j$ is the standard symplectic form on $\mathbb{R}^{2n}$.

The usual Fourier transform F on $\mathbb{R}^{2n}$ being defined by

$$Fa(z) = \left(\frac{1}{2\pi\hbar}\right)^n \int_{\mathbb{R}^{2n}} e^{-\frac{i}{\hbar}z\cdot z'}a(z')dz', \tag{2.13}$$

we have the relationship

$$F_\sigma a(z) = Fa(Jz) \tag{2.14}$$

where J is the standard symplectic matrix (recall that $\sigma(z,z') = Jz\cdot z'$). It follows from (2.14) and the inversion formula

$$F^{-1}a(z) = \left(\frac{1}{2\pi\hbar}\right)^n \int_{\mathbb{R}^{2n}} e^{\frac{i}{\hbar}z\cdot z'}a(z')dz' \tag{2.15}$$

that F_σ is an involution:

$$F_\sigma^{-1} = F_\sigma \Longleftrightarrow F_\sigma^2 = I$$

as follows from the identity

$$F_\sigma(F_\sigma a)(z) = F_\sigma Fa(Jz) = F^2 a(-z) = a(z)$$

where we used the relationship $J^2 = -I_{2n\times 2n}$.

As is the case for the Fourier transform F, the symplectic Fourier transform F_σ is a linear automorphism

$$F_\sigma : \mathcal{S}(\mathbb{R}^{2n}) \longrightarrow \mathcal{S}(\mathbb{R}^{2n})$$

which extends by duality into an automorphism

$$F_\sigma : \mathcal{S}'(\mathbb{R}^{2n}) \longrightarrow \mathcal{S}'(\mathbb{R}^{2n}).$$

In addition to being involutive, F_σ is unitary on $L^2(\mathbb{R}^{2n})$ as follows from (2.14):

$$\|F_\sigma a\|_{L^2(\mathbb{R}^{2n})} = \|a\|_{L^2(\mathbb{R}^{2n})}. \tag{2.16}$$

The symplectic Fourier transform also satisfies the following variants of the Plancherel formula:

$$(F_\sigma a|b)_{L^2(\mathbb{R}^{2n})} = (a|F_\sigma b)_{L^2(\mathbb{R}^{2n})} \tag{2.17}$$

for $(a, b) \in L^2(\mathbb{R}^{2n}) \times L^2(\mathbb{R}^{2n})$. This formula is a straightforward consequence of the fact that F_σ is a unitary involution:

$$\begin{aligned}(F_\sigma a|b)_{L^2(\mathbb{R}^{2n})} &= (F_\sigma^2 a|F_\sigma b)_{L^2(\mathbb{R}^{2n})}\\ &= (a|F_\sigma b)_{L^2(\mathbb{R}^{2n})}.\end{aligned}$$

The symplectic Fourier transform inherits many other properties of the usual Fourier transform; for instance

$$F_\sigma(a * b) = (2\pi\hbar)^n F_\sigma a F_\sigma b \tag{2.18}$$

when defined.

2.4 An analytic relationship between $\widehat{D}(z)$ and $\widehat{R}(z)$

Formula (2.8) is an algebraic relationship between the displacement and reflection operators. Here is a deep analytic relationship between these operators; it will be the key to some important relationships in the next chapter.

Proposition 26. *Let $\psi \in L^2(\mathbb{R}^n)$. We have*

$$\widehat{R}(z_0)\psi(x) = 2^{-n}F_\sigma[\widehat{D}(\cdot)\psi(x)](-z_0). \tag{2.19}$$

This identity also holds for $\psi \in \mathcal{S}'(\mathbb{R}^n)$.

Proof. The proof that follows is somewhat informal, so the rigorous reader can rewrite it by using distributional brackets where necessary. Since $\widehat{R}(z_0)$ and F_σ are continuous automorphisms of $\mathcal{S}'(\mathbb{R}^n)$, it is sufficient to assume that $\psi \in \mathcal{S}(\mathbb{R}^n)$. In view of the explicit expression (2.9), the right-hand side of (2.19) is

$$\begin{aligned} A(x) &= \left(\frac{1}{4\pi\hbar}\right)^n \int_{\mathbb{R}^{2n}} e^{\frac{i}{\hbar}\sigma(z_0,z')}\widehat{D}(z')\psi(x)dz' \\ &= \left(\frac{1}{4\pi\hbar}\right)^n \int_{\mathbb{R}^{2n}} e^{\frac{i}{\hbar}(p_0\cdot x'-p'\cdot x_0+p'\cdot x-\frac{1}{2}p'\cdot x')}\psi(x-x')dz' \\ &= \left(\frac{1}{4\pi\hbar}\right)^n \int_{\mathbb{R}^n}\left(\int_{\mathbb{R}^n} e^{\frac{i}{\hbar}p'\cdot(x-x_0-\frac{1}{2}x')}dp'\right)e^{\frac{i}{\hbar}p_0\cdot x'}\psi(x)dx' \end{aligned}$$

with $z' = (x', p')$. We next note that we have, in the sense of distributions,

$$\int_{\mathbb{R}^{2n}} e^{\frac{i}{\hbar}p'\cdot(x-x_0-\frac{1}{2}x')}dp' = (2\pi\hbar)^n\delta(x-x_0-\frac{1}{2}x'),$$

hence, setting $y = \frac{1}{2}x'$:

$$\begin{aligned} A(x) &= 2^{-n}\int_{\mathbb{R}^n} \delta(x-x_0-\frac{1}{2}x')e^{\frac{i}{\hbar}p_0\cdot x'}\psi(x)dx' \\ &= \int_{\mathbb{R}^n} \delta(y+x_0-x)e^{\frac{2i}{\hbar}p_0\cdot y}\psi(x)dy \\ &= e^{\frac{2i}{\hbar}p_0\cdot(x-x_0)}\psi(-x+2x_0) \end{aligned}$$

which proves (2.9). That the property extends to $\psi \in \mathcal{S}'(\mathbb{R}^n)$ is easily seen; we leave it to the reader to rewrite formula (2.19) in the distributional sense. □

2.5 The Schrödinger representation

Here is a venerable topic that has been known for a long time under various forms and which that physicists see as the quintessence of quantum mechanics. For us, it has mainly the pedagogical advantage of justifying the definition of the Heisenberg displacement operators.

Definition 27. The set $\mathbb{R}^{2n} \times \mathbb{R}_t$ equipped with composition group law

$$(z,t)(z',t) = (z+z',t+t'+\frac{1}{2}\sigma(z,z'))$$

is called the $(2n+1)$-dimensional Heisenberg group H_n.

It is straightforward to verify that H_n indeed is a (noncommutative) group, with unit $(0,0)$ and inverses $(z,t)^{-1} = (-z,-t)$.

Let us now show how the Heisenberg group is linked to the displacement operators $\widehat{D}(z_0)$. Let $\mathcal{U}(L^2(\mathbb{R}^n))$ be the multiplicative group of all unitary operators on $L^2(\mathbb{R}^n)$.

Proposition 28. *The mapping*

$$\widehat{D}(z_0,t_0) : H_n \longrightarrow \mathcal{U}(L^2(\mathbb{R}^n))$$

defined by

$$\widehat{D}(z_0,t_0)\psi(x) = e^{\frac{i}{\hbar}t_0}\widehat{D}(z_0)\psi(x)$$

is a unitary representation of H_n.

Proof. It is clear that $\widehat{D}(z_0,0) = \widehat{D}(z_0)$ and that each operator $\widehat{D}(z_0,t_0)$ is unitary:

$$\|\widehat{D}(z_0,t_0)\psi\|_{L^2} = \|\psi\|_{L^2} \text{ for all } \psi \in L^2(\mathbb{R}^n).$$

On the other hand, by formula (2.6),

$$\widehat{D}(z_0,t_0)\widehat{D}(z_1,t_1) = \widehat{D}(z_0+z_1,t_0+t_1+\frac{1}{2}\sigma(z_0,z_1)), \tag{2.20}$$

hence $\widehat{D}$ is a continuous homomorphism $H_n \longrightarrow \mathcal{U}(L^2(\mathbb{R}^n))$ showing that $\widehat{D}$ is a unitary representation of H_n. □

The Schrödinger representation is irreducible, that is, the only closed subspaces of $L^2(\mathbb{R}^n)$ that are invariant under all $\widehat{D}(z_0,t_0)$ are $\{0\}$ or $L^2(\mathbb{R}^n)$ itself; equivalently, all bounded operators $\widehat{A} \in \mathcal{B}(L^2(\mathbb{R}^n))$ such that

$$\widehat{A}\widehat{D}(z_0,t_0) = \widehat{D}(z_0,t_0)\widehat{A} \text{ for all } (z_0,t_0) \in \mathbb{R}^{2n} \times \mathbb{R}$$

are of the type λI, $\lambda \in \mathbb{C}$. For the sake of completeness, we prove that this is the case. For details, see the section Comments and References below. We will rely on the following classical lemma.

Lemma 29. *Let* $\widehat{A} \in \mathcal{B}(L^2(\mathbb{R}^n))$ *(the bounded operators on* $L^2(\mathbb{R}^n)$*). Let* $T(x_0)$ *be the translation* $x \longmapsto x + x_0$*. If we have* $\widehat{A}T(x_0) = T(x_0)\widehat{A}$ *for all* $x_0 \in \mathbb{R}^n$*, then there exists a function* $a \in L^\infty(\mathbb{R}^n)$ *such that* $F(\widehat{A}\psi) = aF\psi$ *for all* $\psi \in \mathcal{S}(\mathbb{R}^n)$ *(F the Fourier transform on* $\mathbb{R}^n$*).*

Recall that we must show that, if $\widehat{A} \in \mathcal{B}(L^2(\mathbb{R}^n))$ is such that

$$\widehat{A}\widehat{D}(z_0, t_0) = \widehat{D}(z_0, t_0)\widehat{A}$$

for all $(z_0, t_0) \in \mathbb{R}^{2n} \times \mathbb{R}$, then $\widehat{A} = \lambda I$ for some complex constant λ. Choosing $p_0 = 0$ and $t_0 = 0$, we must have, in particular,

$$\widehat{A}T(x_0) = T(x_0)\widehat{A}$$

since $T(x_0) = \widehat{D}(x_0, 0, 0)$. It now suffices to apply Lemma 29.

2.6 Comments and references

The Grossmann–Royer operators were introduced independently in [42] and [64]. For a proof of Lemma 29, we refer to Stein and Weiss [69], Chap. 1, § 3: For a proof of the irreducibility of the Schrödinger representation, we refer to Varadarajan [71] or Wallach [72]; also a chapter in our book [30] (Chap. 10, § 1). For more on the Heisenberg group and a very detailed discussion of the Stone–von Neumann theorem, see § 9.3 of Gröchenig's treatise [40], also Wallach [72] or Folland [28].

3 The cross-Wigner transform

3.1 Definition and properties

3.1.1 Definitions

One of the central objects in quantum harmonic analysis, especially in its phase-space formulation, is the Wigner transform; the latter is a particular case of the cross-Wigner transform:

Definition 30. The sesquilinear mapping

$$W : L^2(\mathbb{R}^n) \times L^2(\mathbb{R}^n) \longrightarrow L^\infty(\mathbb{R}^{2n})$$

defined by

$$W(\psi, \phi) = \left(\frac{1}{\pi\hbar}\right)^n (\widehat{R}(\cdot)\psi|\phi)_{L^2} \tag{3.1}$$

is called the cross-Wigner transform. The function

$$W\psi = W(\psi, \psi) = \left(\frac{1}{\pi\hbar}\right)^n (\widehat{R}(\cdot)\psi|\psi)_{L^2}$$

is called the *Wigner* transform of ψ.

The sesquilinearity of W means that, for all $\psi_1, \psi_2, \phi_1, \phi_2$ in $L^2(\mathbb{R}^n)$ and $\lambda \in \mathbb{C}$, we have

$$\begin{aligned}
W(\psi, \phi_1 + \phi_2) &= W(\psi, \phi_1) + W(\psi, \phi_2)\\
W(\psi_1 + \psi_2, \phi) &= W(\psi_1, \phi) + W(\psi_2, \phi)\\
W(\lambda\psi, \phi) &= \lambda W(\psi, \phi)\\
W(\psi, \lambda\phi) &= \bar{\lambda} W(\psi, \phi).
\end{aligned}$$

Observe that

$$\overline{W(\psi, \phi)} = W(\phi, \psi) \quad \text{and} \quad W\psi(z) \in \mathbb{R},$$

that is, $W\psi$ is a real function.

That this definition makes sense follows from the Cauchy–Schwarz inequality that tells us (taking the unitarity of the reflection operator into account) that

$$|W(\phi, \psi)(z)| \leq \left(\frac{1}{\pi\hbar}\right)^n \|\psi\|_{L^2}^2 \|\phi\|_{L^2}^2,$$

https://doi.org/10.1515/9783110722772-003

hence $W(\phi,\psi) \in L^{\infty}(\mathbb{R}^{2n})$. We leave it to the reader to check, using the Leibniz rule for differentiation,

$$W : \mathcal{S}(\mathbb{R}^n) \times \mathcal{S}(\mathbb{R}^n) \longrightarrow \mathcal{S}(\mathbb{R}^{2n}). \tag{3.2}$$

The reader will perhaps object that this is not the definition of the Wigner transform he has been taught in college that is

$$W\psi(z) = \left(\frac{1}{2\pi\hbar}\right)^n \int_{\mathbb{R}^n} e^{-\frac{i}{\hbar}p\cdot y}\psi\left(x + \frac{1}{2}y\right)\overline{\psi\left(x - \frac{1}{2}y\right)}dy. \tag{3.3}$$

It turns out that both definitions are equivalent; in fact, we have more generally

$$W(\psi,\phi)(z) = \left(\frac{1}{2\pi\hbar}\right)^n \int_{\mathbb{R}^n} e^{-\frac{i}{\hbar}p\cdot y}\psi\left(x + \frac{1}{2}y\right)\overline{\phi\left(x - \frac{1}{2}y\right)}dy. \tag{3.4}$$

To prove that both definitions coincide, we begin by noting that, by definition of the reflection operator $\widehat{R}(z_0)$, we have

$$(\widehat{R}(z_0)\psi|\phi)_{L^2} = \int_{\mathbb{R}^n} e^{\frac{2i}{\hbar}p_0\cdot(x-x_0)}\psi(2x_0 - x)\overline{\phi(x)}dx,$$

and hence, setting $y = 2(x_0 - x)$,

$$(\widehat{R}(z_0)\psi|\phi)_{L^2} = 2^{-n} \int_{\mathbb{R}^n} e^{-\frac{i}{\hbar}p_0\cdot y}\psi\left(x_0 + \frac{1}{2}y\right)\overline{\phi\left(x_0 - \frac{1}{2}y\right)}dy$$

that is

$$(\widehat{R}(z_0)\psi|\phi)_{L^2} = (\pi\hbar)^n W(\psi,\phi)(z_0)$$

proving (3.3) in view of (3.1).

3.1.2 Elementary properties

We begin by showing a normalization result that $W\psi$ is absolutely integrable if ψ is square integrable. This result is important for the study of the probabilistic properties of the Wigner transform:

Proposition 31. *Let $\psi,\phi \in L^2(\mathbb{R}^n)$, and suppose that $W(\psi,\phi) \in L^1(\mathbb{R}^{2n})$. Then*

$$\int_{\mathbb{R}^{2n}} W(\psi,\phi)(z)dz = (\psi|\phi)_{L^2}. \tag{3.5}$$

In particular, if $\psi \in L^2(\mathbb{R}^n)$ and $W\psi \in L^1(\mathbb{R}^{2n})$, then $W\psi \in L^1(\mathbb{R}^{2n})$ and

$$\int_{\mathbb{R}^{2n}} W\psi(z)dz = \|\psi\|^2_{L^2}. \tag{3.6}$$

Proof. Assume that $\psi, \phi \in \mathcal{S}(\mathbb{R}^n)$. We begin by noting that by definition of the reflection operator $\widehat{R}(z_0)$ we have

$$\begin{aligned}\int_{\mathbb{R}^{2n}} \widehat{R}(z_0)\psi(x)dz_0 &= \int_{\mathbb{R}^{2n}} e^{\frac{2i}{\hbar}p_0\cdot(x-x_0)}\psi(2x_0-x)dp_0dx_0 \\ &= \int_{\mathbb{R}^n} e^{\frac{2i}{\hbar}p_0\cdot(x-x_0)}\Big(\int_{\mathbb{R}^n}\psi(2x_0-x)dx_0\Big)dp_0 \\ &= (\pi\hbar)^n\int_{\mathbb{R}^n}\delta(x-x_0)\psi(2x_0-x)dx_0 \\ &= (\pi\hbar)^n\psi(x).\end{aligned}$$

It follows that

$$\begin{aligned}\int_{\mathbb{R}^{2n}} W(\psi,\phi)(z_0)dz_0 &= \Big(\frac{1}{\pi\hbar}\Big)^n\Big(\int_{\mathbb{R}^{2n}}\widehat{R}(z_0)\psi(x)\overline{\phi(x)}dz_0\Big)dx \\ &= \Big(\frac{1}{\pi\hbar}\Big)^n\int_{\mathbb{R}^n}\Big(\int_{\mathbb{R}^{2n}}\widehat{R}(z_0)\psi(x)dz_0\Big)\overline{\phi(x)}dx \\ &= \int_{\mathbb{R}^n}\psi(x)\overline{\phi(x)}dx\end{aligned}$$

which is formula (3.5). □

An immediate consequence of the definition of the cross-Wigner transform is that $W(\psi,\phi)$ is bounded and continuous if ψ and ϕ are both square integrable:

Proposition 32. *Let $(\psi,\phi) \in L^2(\mathbb{R}^n)\times L^2(\mathbb{R}^n)$. The function $z \longmapsto W(\psi,\phi)(z)$ is bounded and continuous on $\mathbb{R}^{2n}$, that is*

$$W(\psi,\phi) \in L^\infty(\mathbb{R}^{2n}) \cap C^0(\mathbb{R}^{2n}).$$

Proof. We have already shown that $W(\psi,\phi) \in L^\infty(\mathbb{R}^{2n})$. Let us prove the continuity of $W(\psi,\phi)$. It suffices to do so for $\psi,\phi \in \mathcal{S}(\mathbb{R}^n)$ since $\mathcal{S}(\mathbb{R}^n)$ is dense in $L^2(\mathbb{R}^n)$. Set

$$f(z,u) = \Big(\frac{1}{\pi\hbar}\Big)^n\widehat{R}(z)\psi(u)\overline{\phi(u)};$$

the function $z \longmapsto f(z,u)$ is continuous, and we have $|f(z,u)| \leq g(u)$ where $g(u) \leq C|\phi(u)|$ for some constant $C > 0$ (because $u \longmapsto R(z)\psi(u)$ is bounded). The continuity of $W(\psi,\phi)$ now follows applying Lebesgue's dominated convergence theorem. □

Talking about continuity, we have the following very general result:

Proposition 33. *Let $\psi,\phi \in \mathcal{S}'(\mathbb{R}^n)$. The function $z \longmapsto W(\psi,\phi)(z)$ is continuous on $\mathbb{R}^{2n}$.*

Proof. Assume first that $\psi, \phi \in \mathcal{S}(\mathbb{R}^n)$. Let us show that

$$\lim_{z \to z_0} (W(\psi, \phi)(z) - W(\psi, \phi)(z_0)) = 0. \tag{3.7}$$

We have previously shown above the mapping $z \longmapsto \widehat{R}(z)\psi$ is strongly continuous on $\mathbb{R}^{2n}$; we have thus in particular

$$\lim_{z \to z_0} \|\widehat{R}(z)\psi - \widehat{R}(z_0)\psi\|_\infty = 0.$$

In view of the first definition of $W(\psi, \phi)(z)$, we have

$$W(\psi, \phi)(z) - W(\psi, \phi)(z_0) = \left(\frac{1}{\pi\hbar}\right)^n \int_{\mathbb{R}^n} (\widehat{R}(z)\psi(x') - \widehat{R}(z_0)\psi(x'))\overline{\phi(x')}dx'$$

hence

$$|W(\psi, \phi)(z) - W(\psi, \phi)(z_0)| \leq \left(\frac{1}{\pi\hbar}\right)^n \|\widehat{R}(z)\psi - \widehat{R}(z_0)\psi\|_\infty \|\phi\|_1$$

and (3.7). The case $\psi, \phi \in \mathcal{S}'(\mathbb{R}^n)$ follows by duality. □

3.1.3 Hudson's theorem

R. Hudson's celebrated theorem says that the Wigner transform of $\psi \in L^2(\mathbb{R}^n)$ can be positive if and only if ψ is a non-degenerate Gaussian. In our notation and terminology, this result can be restated as follows:

Proposition 34. *Let $\psi \in L^2(\mathbb{R}^n)$. We have $W\psi > 0$ if and only if $\psi = \widehat{D}(z_0)\psi_M$ for some $z_0 \in \mathbb{R}^{2n}$.*

We omit the proof of Hudson's theorem; see the subsequent comments and references section.

We mention the following refinement of Hudson's theorem due to J. Toft:

Proposition 35. *Let $\psi, \phi \in L^2(\mathbb{R}^n)$. We have $W(\psi, \phi) > 0$ if and only if $\psi = T(z_0)\psi_M$ and $\phi = \lambda\psi$ for some $\lambda > 0$.*

Thus, we cannot expect the cross-Wigner transform to be positive for any distinct state.

3.1.4 Displacing cross-Wigner transforms

The following important result describes the behavior of the cross-Wigner transform under displacements:

Proposition 36. *For every pair (ψ, ϕ) of functions in $\mathcal{S}'(\mathbb{R}^n)$, we have*

$$W(\widehat{D}(z_0)\psi, \widehat{D}(z_1)\phi)(z) = e^{-\frac{i}{\hbar}[\sigma(z,z_0-z_1)+\frac{1}{2}\sigma(z_0,z_1)]}W(\psi,\phi)(z-\langle z\rangle) \tag{3.8}$$

where $\langle z\rangle = \frac{1}{2}(z_0+z_1)$. In particular:

$$W(\widehat{D}(z_0)\psi, \widehat{D}(z_0)\phi)(z) = W(\psi,\phi)(z-z_0) \tag{3.9}$$

$$W(\widehat{D}(z_0)\psi)(z) = W\psi(z-z_0) \tag{3.10}$$

$$W(\widehat{D}(z_0)\psi, \phi)(z) = e^{-\frac{i}{\hbar}\sigma(z,z_0)}W(\psi,\phi)\left(z-\frac{1}{2}z_0\right). \tag{3.11}$$

Proof. To prove formula (3.8), it is sufficient to assume that $\psi, \phi \in \mathcal{S}(\mathbb{R}^n)$. We will set $\langle x\rangle = \frac{1}{2}(x_0+x_1)$ and $\langle p\rangle = \frac{1}{2}(p_0+p_1)$. By definition of the Heisenberg–Weyl operator, we have

$$\widehat{D}(z_0)\psi\left(x+\frac{1}{2}y\right) = e^{\frac{i}{\hbar}[p_0\cdot(x+\frac{1}{2}y)-\frac{1}{2}p_0\cdot x_0]}\psi\left(x-x_0+\frac{1}{2}y\right)$$

$$\widehat{D}(z_1)\phi\left(x-\frac{1}{2}y\right) = e^{\frac{i}{\hbar}[p_1\cdot(x-\frac{1}{2}y)-\frac{1}{2}p_1\cdot x_1]}\phi\left(x-x_1-\frac{1}{2}y\right),$$

and hence

$$\widehat{D}(z_0)\psi\left(x+\frac{1}{2}y\right)\overline{\widehat{D}(z_1)\phi\left(x-\frac{1}{2}y\right)} = e^{\frac{i}{\hbar}\delta(z_0,z_1)}e^{\frac{i}{\hbar}\langle p\rangle\cdot y}\psi\left(x-x_0+\frac{1}{2}y\right)\overline{\phi\left(x-x_1-\frac{1}{2}y\right)}$$

with

$$\delta(z_0,z_1) = (p_0-p_1)\cdot x - \frac{1}{2}(p_0\cdot x_0 - p_1\cdot x_1).$$

It follows that we have

$$W(\widehat{D}(z_0)\psi, \widehat{D}(z_1)\phi)(z) = \left(\frac{1}{2\pi\hbar}\right)^n e^{\frac{i}{\hbar}\delta(z_0,z_1)} \times \int_{\mathbb{R}^n} e^{-\frac{i}{\hbar}(p-\langle p\rangle)\cdot y}\psi\left(x-x_0+\frac{1}{2}y\right)\overline{\phi\left(x-x_0-\frac{1}{2}y\right)}dy.$$

Performing the change of variables $y' = x_1 - x_0 + y$ in the integral, this equality becomes

$$W(\widehat{D}(z_0)\psi, \widehat{D}(z_1)\phi)(z) = \left(\frac{1}{2\pi\hbar}\right)^n e^{\frac{i}{\hbar}\Delta} \times \int_{\mathbb{R}^n} e^{-\frac{i}{\hbar}(p-\langle p\rangle)\cdot y}\psi\left(x-\langle x\rangle+\frac{1}{2}y\right)\overline{\phi\left(x-\langle x\rangle-\frac{1}{2}y\right)}dy$$

where the phase Δ is given by

$$\Delta = (p_0-p_1)\cdot x - (x_0-x_1)\cdot p + \frac{1}{2}(p_1\cdot x_0 - p_0\cdot x_1)$$

$$= -\sigma(z, z_0 - z_1) - \frac{1}{2}\sigma(z_0, z_1),$$

hence formula (3.8). The formulas (3.9), (3.10), and (3.11) immediately follow from (3.8). □

3.2 The Moyal identity

3.2.1 Statement and proof

The main result of this section is formula (3.12) whose importance cannot be overestimated.

Proposition 37. *The cross-Wigner transform satisfies the "Moyal identity"*

$$(W(\psi,\phi)|W(\psi',\phi'))_{L^2(\mathbb{R}^{2n})} = \left(\frac{1}{2\pi\hbar}\right)^n (\psi|\psi')_{L^2}\overline{(\phi|\phi')_{L^2}} \tag{3.12}$$

for all $(\psi,\phi) \in L^2(\mathbb{R}^n) \times L^2(\mathbb{R}^n)$ *and* $(\psi',\phi') \in L^2(\mathbb{R}^n) \times L^2(\mathbb{R}^n)$. *In particular*

$$\|W(\psi,\phi)\|_{L^2(\mathbb{R}^{2n})} = \left(\frac{1}{2\pi\hbar}\right)^n \|\psi\|_{L^2}\|\phi\|_{L^2} \tag{3.13}$$

and

$$\|W\psi\|_{L^2(\mathbb{R}^{2n})} = \left(\frac{1}{2\pi\hbar}\right)^{n/2} \|\psi\|^2_{L^2}. \tag{3.14}$$

Proof. Let us set

$$A = (2\pi\hbar)^{2n}(W(\psi,\phi)|W(\psi',\phi'))_{L^2(\mathbb{R}^{2n})}.$$

In view of the integral definition (3.4) of the cross-Wigner transform, we have

$$A = \int_{\mathbb{R}^{4n}} e^{-\frac{i}{\hbar}p\cdot(y-y')}\psi\left(x+\frac{1}{2}y\right)\overline{\psi'\left(x+\frac{1}{2}y'\right)} \\ \times \overline{\phi\left(x-\frac{1}{2}y\right)}\phi'\left(x-\frac{1}{2}y'\right)dydy'dxdp.$$

The integral in p can be viewed as an inverse Fourier transform, yielding

$$\int_{\mathbb{R}^n} e^{-\frac{i}{\hbar}p\cdot(y-y')}dp = (2\pi\hbar)^n\delta(y-y'),$$

and hence

$$A = (2\pi\hbar)^n \int_{\mathbb{R}^{3n}} \psi\left(x+\frac{1}{2}y\right)\overline{\psi'\left(x-\frac{1}{2}y\right)\phi\left(x+\frac{1}{2}y\right)}\phi'\left(x-\frac{1}{2}y\right)dydy'dx.$$

Setting $x' = x + \frac{1}{2}y$ and $y' = x - \frac{1}{2}y$, we have $dx'dy' = dxdy$, and hence

$$A = (2\pi\hbar)^n \Big(\int_{\mathbb{R}^n} \psi(x')\overline{\psi'(x')}dx' \Big)\Big(\int_{\mathbb{R}^n} \overline{\phi(y')}\phi'(y')dy' \Big)$$

which proves Moyal's identity (3.12). Formula (3.14) immediately follows. □

Notice that the Moyal identity can be written, using distributional brackets, as

$$\langle W(\psi,\phi), W(\psi',\phi')\rangle = \left(\frac{1}{2\pi\hbar}\right)^n \langle \psi, \overline{\psi'}\rangle\langle \phi, \overline{\phi'}\rangle. \tag{3.15}$$

3.2.2 An extension result for the cross-Wigner transform

The Moyal identity implies the following continuity result for the cross-Wigner transform:

Proposition 38. *The cross-Wigner transform is a continuous mapping*

$$W : L^2(\mathbb{R}^n) \times L^2(\mathbb{R}^n) \longrightarrow L^2(\mathbb{R}^{2n}). \tag{3.16}$$

Proof. Choosing $\psi = \psi'$ and $\phi = \phi'$ in Moyal's identity (3.12), we get

$$\|W(\psi,\phi)\|_{L^2(\mathbb{R}^{2n})} = \left(\frac{1}{2\pi\hbar}\right)^{n/2} \|\psi\|_{L^2}\|\phi\|_{L^2}$$

hence the sesquilinear mapping $(\psi,\phi) \longmapsto W(\psi,\phi)$ is a continuous mapping $L^2(\mathbb{R}^n) \times L^2(\mathbb{R}^n) \longrightarrow L^2(\mathbb{R}^{2n})$. □

Recall that ϕ_0 is the standard Gaussian on $\mathbb{R}^n$ defined by

$$\phi_0(x) = (\pi\hbar)^{-n}e^{-|x|^2/2\hbar}.$$

Corollary 39. *Let $\phi_0 \in L^2(\mathbb{R}^n)$ such that $\|\phi_0\|_{L^2} \neq 0$. The set $\{\widehat{D}(z)\phi_0 : z \in \mathbb{R}^{2n}\}$ spans a dense subspace X of $L^2(\mathbb{R}^n)$.*

Proof. Let $\psi \in L^2(\mathbb{R}^n)$. Let us prove that $(\psi|\widehat{D}(z)\phi_0)_{L^2} = 0$ for all $z \in \mathbb{R}^n$ implies that $\psi = 0$; the result will follow. We have

$$(\psi|\widehat{D}(z)\phi_0)_{L^2} = (\pi\hbar)^n W(\psi,\phi_0)(z),$$

hence $(\psi|\widehat{D}(z)\phi_0)_{L^2} = 0$ for all z is equivalent to $W(\psi,\phi_0)(z) = 0$ for all z. In view of Moyal's identity, we then have $\|\psi\|_{L^2}\|\phi_0\|_{L^2} = 0$, and hence $\psi = 0$. □

Note that this result says that, if $W(\psi,\phi) = 0$ for all ϕ then $\psi = 0$. A similar result holds for the cross-ambiguity function: If $A(\psi,\phi) = 0$ for all ϕ, then $\psi = 0$.

3.2.3 A reconstruction result

Moyal's identity allows us in addition to prove an inversion formula for the Wigner function.

Proposition 40. *Let* $(\phi, \gamma) \in L^2(\mathbb{R}^n) \times L^2(\mathbb{R}^n)$ *be such that* $(\gamma|\phi)_{L^2} \neq 0$. *For every* $\psi \in \mathcal{S}'(\mathbb{R}^n)$, *we have*

$$\psi = \frac{2^n}{(\gamma|\phi)_{L^2}} \int_{\mathbb{R}^{2n}} W(\psi, \phi)(z)\widehat{R}(z)\gamma dz \tag{3.17}$$

almost everywhere.

Proof. Let us denote by f the right-hand side of (3.17). This function is well-defined since $W(\psi, \phi) \in L^2(\mathbb{R}^{2n})$ in view of Moyal's identity. For any $\alpha \in \mathcal{S}(\mathbb{R}^n)$, we have

$$(f|\alpha)_{L^2} = \frac{2^n}{(\gamma|\phi)_{L^2}} \int_{\mathbb{R}^{2n}} W(\psi, \phi)(z)(\widehat{R}(z)\gamma|\alpha)_{L^2} dz.$$

Recalling that by definition

$$W(\psi, \phi)(z) = \left(\frac{1}{\pi\hbar}\right)^n (\widehat{R}(z)\psi|\phi)_{L^2},$$

we thus have the sequence of equalities

$$\begin{aligned}(f|\alpha) &= \frac{(2\pi\hbar)^n}{(\gamma|\phi)_{L^2}} \int_{\mathbb{R}^{2n}} W(\psi, \phi)(z)W(\gamma, \alpha)dz \\ &= \frac{(2\pi\hbar)^n}{(\gamma|\phi)_{L^2}} \int_{\mathbb{R}^{2n}} W(\psi, \phi)(z)\overline{W(\alpha, \gamma)}dz \\ &= \frac{(2\pi\hbar)^n}{(\gamma|\phi)_{L^2}} (W(\psi, \phi)|W(\alpha, \gamma))_{L^2}.\end{aligned}$$

Applying Moyal's identity (3.12) to the product $(W(\psi, \phi)|W(\alpha, \gamma))_{L^2}$, we get

$$(f|\alpha)_{L^2} = \frac{1}{(\gamma|\phi)} (\psi|\alpha)_{L^2}\overline{(\phi|\gamma)_{L^2}} = (\psi|\alpha)_{L^2}.$$

Since this identity holds for all $\alpha \in \mathcal{S}(\mathbb{R}^n)$., we have $f = \psi$ almost everywhere, hence (3.17). □

3.3 The cross-ambiguity function

3.3.1 Definition

Closely related to the cross-Wigner transform is the ambiguity function since they are symplectic Fourier transforms of each other. We begin by giving a definition in terms of the Heisenberg displacement operator.

Definition 41. The cross-ambiguity function on $\mathbb{R}^n$ is the sesquilinear mapping

$$A : L^2(\mathbb{R}^n) \times L^2(\mathbb{R}^n) \longrightarrow L^\infty(\mathbb{R}^{2n})$$

defined by

$$A(\psi,\phi)(z) = \left(\frac{1}{2\pi\hbar}\right)^n (\psi|\widehat{D}(z)\phi)_{L^2}. \tag{3.18}$$

The function $A\psi = A(\psi,\psi)$ is called the auto-ambiguity function:

$$A\psi(z) = \left(\frac{1}{2\pi\hbar}\right)^n (\psi|\widehat{D}(z)\psi)_{L^2}. \tag{3.19}$$

The cross-ambiguity function has the property that (cf. (3.2))

$$A : \mathcal{S}(\mathbb{R}^n) \times \mathcal{S}(\mathbb{R}^n) \longrightarrow \mathcal{S}(\mathbb{R}^{2n}). \tag{3.20}$$

Using the definition of $\widehat{D}(z)$, it is easily seen that the cross-ambiguity function of $\psi,\phi \in L^2(\mathbb{R}^n)$ is explicitly given by the formula

$$A(\psi,\phi)(z) = \left(\frac{1}{2\pi\hbar}\right)^n \int_{\mathbb{R}^n} e^{-\frac{i}{\hbar}p\cdot y}\psi\left(y+\frac{1}{2}x\right)\overline{\phi\left(y-\frac{1}{2}x\right)}dy\,; \tag{3.21}$$

as for the cross-Wigner, transform it is immediately checked, using the Cauchy–Schwarz inequality, that

$$|A(\psi,\phi)(z)| \leq \left(\frac{1}{2\pi\hbar}\right)^n \|\psi\|_{L^2}\|\phi\|_{L^2}.$$

3.3.2 Relations with the cross-Wigner transform

There are two ways to relate the cross-Wigner and ambiguity functions. One is algebraic, and the other is analytical.

Proposition 42. *We have*

$$A(\psi,\phi)(z) = 2^{-n}W(\psi,\phi^\vee)\left(\frac{1}{2}z\right) \tag{3.22}$$

where $\phi^\vee(x) = \phi(-x)$.

Proof. By definition of $W(\psi,\phi)$, we have

$$W(\psi,\phi)\left(\frac{1}{2}z\right) = \left(\frac{1}{2\pi\hbar}\right)^n \int_{\mathbb{R}^n} e^{-\frac{i}{2\hbar}p\cdot y}\psi\left(\frac{1}{2}x+\frac{1}{2}y\right)\overline{\phi\left(\frac{1}{2}x-\frac{1}{2}y\right)}dy\,;$$

setting $x' = \frac{1}{2}y$, this is

$$W(\psi,\phi^\vee)\left(\frac{1}{2}z\right) = \left(\frac{1}{\pi\hbar}\right)^n \int_{\mathbb{R}^n} e^{-\frac{i}{\hbar}p\cdot x'}\psi\left(x'+\frac{1}{2}x\right)\overline{\phi^\vee\left(x'-\frac{1}{2}x\right)}dx',$$

hence (3.22) in view of (3.21). □

Many of the properties of the ambiguity function follow from the previous above; for instance, we have the following analogue of Proposition 33:

Proposition 43. *Let $\psi,\phi \in \mathcal{S}(\mathbb{R}^n)$. The mapping $z \longmapsto A(\psi,\phi)(z)$ is continuous on $\mathbb{R}^{2n}$.*

The relationship (3.22) is purely algebraic; the following result is analytical:

Proposition 44. *Let ψ and ϕ be in $\mathcal{S}'(\mathbb{R}^n)$. Let F_σ be the symplectic Fourier transform. We have*

$$A(\psi,\phi) = F_\sigma(W(\psi,\phi)), \quad W(\psi,\phi) = F_\sigma(A(\psi,\phi)). \tag{3.23}$$

In particular, $A\psi = F_\sigma(W\psi)$, and $W\psi = F_\sigma(A\psi)$.

Proof. It is sufficient to show that $A(\psi,\phi) = F_\sigma W(\psi,\phi)$: Since the symplectic Fourier transform is involutive, both formulas (3.23) are equivalent. It is moreover sufficient to assume that ψ and ϕ are in the Schwartz space $\mathcal{S}(\mathbb{R}^n)$. Setting

$$f = (2\pi\hbar)^{2n}F_\sigma W(\psi,\phi),$$

we have, by definition of F_σ and $W(\psi,\phi)$,

$$\begin{aligned} f(z) &= \int_{\mathbb{R}^{3n}} e^{-\frac{i}{\hbar}[\sigma(z,z')+p'\cdot(y-x)]}\psi\left(x'+\frac{1}{2}y\right)\overline{\phi\left(x'-\frac{1}{2}y\right)}dp'dx'dy \\ &= \int_{\mathbb{R}^{3n}} e^{-\frac{i}{\hbar}p'\cdot(y-x)}e^{-\frac{i}{\hbar}p\cdot x'}\psi\left(x'+\frac{1}{2}y\right)\overline{\phi\left(x'-\frac{1}{2}y\right)}dp'dx'dy. \end{aligned}$$

Using the Fourier inversion formula, informally written as

$$\int_{\mathbb{R}^n} e^{-\frac{i}{\hbar}p'\cdot(y-x)}dp' = (2\pi\hbar)^n\delta(x-y),$$

we have

$$f(z) = (2\pi\hbar)^n \int_{\mathbb{R}^{2n}} \delta(x-y)e^{-\frac{i}{\hbar}p\cdot x'}\psi\left(x'+\frac{1}{2}y\right)\overline{\phi\left(x'-\frac{1}{2}y\right)}dx'dy$$
$$= (2\pi\hbar)^n \int_{\mathbb{R}^n} e^{-\frac{i}{\hbar}p\cdot x'}\psi\left(x'+\frac{1}{2}x\right)\overline{\phi\left(x'-\frac{1}{2}x\right)}dx',$$

hence $f = A(\psi,\phi)$, which proves our assertion. □

3.3.3 Moyal identity for the cross-ambiguity function

This result has as an immediate consequence the fact that the ambiguity function also satisfies an Moyal identity:

Proposition 45. *The cross-ambiguity function satisfies the Moyal identity*

$$(A(\psi,\phi)|A(\psi',\phi'))_{L^2(\mathbb{R}^{2n})} = \left(\frac{1}{2\pi\hbar}\right)^n (\psi|\psi')_{L^2}\overline{(\phi|\phi')_{L^2}} \tag{3.24}$$

for all $\psi,\phi \in L^2(\mathbb{R}^n)$. *In particular,*

$$\|A\psi\|_{L^2(\mathbb{R}^{2n})} = \left(\frac{1}{2\pi\hbar}\right)^{n/2} \|\psi\|_{L^2}. \tag{3.25}$$

Proof. We have, taking the Moyal identity in Proposition 37 and the formula

$$A(\psi,\phi) = F_\sigma(W(\psi,\phi))$$

into account,

$$(A(\psi,\phi)|A(\psi',\phi'))_{L^2(\mathbb{R}^{2n})} = F_\sigma(W(\psi,\phi))|F_\sigma(W(\psi',\phi'))_{L^2(\mathbb{R}^{2n})},$$

hence formula (3.24) since F_σ is a unitary operator on $L^2(\mathbb{R}^{2n})$. □

The Moyal identity has the following interesting application to the convolution of Wigner transforms:

Proposition 46. *Let* $(\psi,\phi) \in L^2(\mathbb{R}^n) \times L^2(\mathbb{R}^n)$. *We have*

$$W\psi * W\phi = A(\psi^\vee,\phi)|^2 \tag{3.26}$$

where $\psi^\vee(x) = \psi(-x)$ *and* $A(\cdot,\cdot)$ *is the cross-ambiguity function.*

Proof. We have, since $W\phi$ is a real function,

$$(W\psi * W\phi)(z) = \int_{\mathbb{R}^{2n}} W\psi(z - z_0) W\phi(z_0) dz_0$$
$$= \int_{\mathbb{R}^{2n}} W(\widehat{D}(z)\psi^{\vee})(z_0) W\phi(z_0) dz_0,$$

hence, by the Moyal identity,

$$(W\psi * W\phi)(z) = \left(\frac{1}{2\pi\hbar}\right)^n |(\widehat{D}(z)\psi^{\vee}|\phi)_{L^2}|^2.$$

Formula (3.26) follows in view of the definition (3.18) of the cross-ambiguity transform. □

3.4 Dependence of the Wigner transform on *h*

Let us now address the following question: for given $\psi \in L^2(\mathbb{R}^n)$, can we find ϕ such that $W_\eta\phi = W\psi$ for $\eta \neq \hbar$? The answer is negative. We will use the F_η Fourier transform, obtained from F by replacing $\hbar$ with η:

$$F_\eta\psi(p) = \left(\frac{1}{2\pi\eta}\right)^{n/2} \int_{\mathbb{R}^n} e^{-\frac{i}{\eta}p\cdot x}\psi(x)dx,$$

and, similarly, denote by W_η the η-dependent Wigner transform:

$$W_\eta(\psi,\phi)(z) = \left(\frac{1}{2\pi\eta}\right)^n \int_{\mathbb{R}^n} e^{-\frac{i}{\eta}p\cdot y}\psi\left(x + \frac{1}{2}y\right)\overline{\phi\left(x - \frac{1}{2}y\right)}dy$$

and $W_\eta\psi = W_\eta(\psi,\psi)$. When $\eta = \hbar$, we will write F, $W\psi$, etc. as usual.

Proposition 47. *Let $\psi \in L^1(\mathbb{R}^n) \cap L^2(\mathbb{R}^n)$ ($\psi \neq 0$) and $\eta > 0$. Then:* (i) *There does not exist any $\phi \in L^1(\mathbb{R}^n) \cap L^2(\mathbb{R}^n)$ such that $W_\eta\phi = W\psi$ if $\eta \neq \hbar$;* (ii) *Assume that there exists a sequence $(\psi_j)_j$ of functions $\psi_j \in L^1(\mathbb{R}^n) \cap L^2(\mathbb{R}^n)$, with $\|\psi_j\|_2 = 1$ for every j, and a sequence $(\alpha_j)_j \in \ell^1(\mathbb{N})$ of nonnegative real numbers such that*

$$W\psi = \sum_j \alpha_j W_\eta\psi_j. \tag{3.27}$$

Then we must have $\eta \leq \hbar$.

Proof. (i) Assume that $W_\eta\phi = W\psi$; then, since $\psi \in L^1(\mathbb{R}^n) \cap L^2(\mathbb{R}^n)$, the marginal property (3.29)

$$|F_\eta\phi(p)|^2 = \int_{\mathbb{R}^n} W_\eta\phi(x,p)dx$$

applies, and we thus have since $W_\eta\phi = W\psi$,

$$|F_\eta\phi(p)|^2 = \int_{\mathbb{R}^n} W\psi(x,p)dx = |F_\hbar\psi(p)|^2,$$

hence, by Parseval's equality, ϕ and ψ must have the same L^2-norm: $\|\phi\| = \|\psi\|$. On the other hand, using the Moyal identity, the equality $W\psi = W_\eta\phi$ implies that

$$\int_{\mathbb{R}^{2n}} W\psi(z)^2 dz = \left(\frac{1}{2\pi\hbar}\right)^n \|\psi\|^4$$

$$\int_{\mathbb{R}^{2n}} W_\eta\phi(z)^2 dz = \left(\frac{1}{2\pi\eta}\right)^n \|\phi\|^4,$$

hence we must have $\eta = \hbar$. (ii) Observe that the series is absolutely convergent in $L^2(\mathbb{R}^{2n})$. Proceeding in the same way, we get, using again the first marginal property and Parseval's equality,

$$\|\psi\|^2 = \sum_j \alpha_j \|\psi_j\|^2 = \sum_j \alpha_j. \tag{3.28}$$

On the other hand, squaring $W\psi$ we get

$$(W\psi)^2 = \sum_{j,k} \alpha_j \alpha_k W_\eta \psi W_\eta \psi_k,$$

hence, integrating and using the Moyal identity for $(W\psi)^2$ and $W_\eta\phi_j W_\eta\phi_k$, together with the Cauchy–Schwarz inequality we get, since the α_j are nonnegative,

$$\begin{aligned}\left(\frac{1}{2\pi\hbar}\right)^n \|\psi\|^4 &= \left(\frac{1}{2\pi\eta}\right)^n \sum_{j,k} \alpha_j\alpha_k |(\psi_j|\psi_k)|^2 \\ &\leq \left(\frac{1}{2\pi\eta}\right)^n \sum_{j,k} \alpha_j\alpha_k \|\psi_j\|^2 \|\psi_k\|^2 \\ &= \left(\frac{1}{2\pi\eta}\right)^n \left(\sum_j \alpha_j\right)^2.\end{aligned}$$

Taking the equality (3.28) into account

$$\left(\frac{1}{2\pi\hbar}\right)^n \|\psi\|^4 \leq \left(\frac{1}{2\pi\eta}\right)^n \|\psi\|^4$$

which implies $\eta \leq \hbar$ as claimed. □

3.5 Statistical interpretation of the Wigner transform

The Wigner transform $W\psi$ of $\psi \in L^2(\mathbb{R}^n)$ is a real function. Suppose that $\|\psi\|_{L^2} = 1$. Then, by formula (3.6) in Proposition 31,

$$\int_{\mathbb{R}^{2n}} W\psi(z)dz = 1.$$

This formula, together with the fact that $W\psi$ always is real, suggests that the Wigner transform might be viewed as a "quasi-probability distribution" on $\mathbb{R}^{2n}$. This idea is actually confirmed by the following result, which shows that $W\psi(z)$ satisfies the expected marginal properties.

Proposition 48. *Assume that $\psi \in L^1(\mathbb{R}^n) \cap L^2(\mathbb{R}^n)$ and $F\psi \in L^1(\mathbb{R}^n) \cap L^2(\mathbb{R}^n)$. Then*

$$\int_{\mathbb{R}^n} W\psi(z)dp = |\psi(x)|^2, \quad \int_{\mathbb{R}^n} W\psi(z)dx = |F\psi(p)|^2. \tag{3.29}$$

Proof. Let us prove that, more generally,

$$\int_{\mathbb{R}^n} W(\psi,\phi)(z)dp = \psi(x)\overline{\phi(x)} \tag{3.30}$$

$$\int_{\mathbb{R}^n} W(\psi,\phi)(z)dx = F\psi(p)\overline{F\phi(p)} \tag{3.31}$$

when ψ and ϕ both satisfy the stated conditions. Formulas (3.29) will follow taking $\psi = \phi$. Writing the Fourier inversion formula in distributional form as

$$\int_{\mathbb{R}^n} e^{-\frac{i}{\hbar}p\cdot y}dp = (2\pi\hbar)^n\delta(y),$$

we have

$$\begin{aligned}\int_{\mathbb{R}^n} W(\psi,\phi)(z)dp &= \int_{\mathbb{R}^n} \delta(y)\psi\left(x+\frac{1}{2}y\right)\overline{\phi\left(x-\frac{1}{2}y\right)}dy \\ &= \int_{\mathbb{R}^n} \delta(y)\psi(x)\overline{\phi(x)}dy \\ &= \psi(x)\overline{\phi(x)}\end{aligned}$$

as claimed. This proves formula (3.30). Let us prove (3.31). Making the unitary change of variables $(x,y) \longmapsto (x+\frac{1}{2}y, x-\frac{1}{2}y)$ on the right-hand of the equality

$$\int_{\mathbb{R}^n} W\psi(z)dx = \left(\frac{1}{2\pi\hbar}\right)^n \int_{\mathbb{R}^{2n}} e^{-\frac{i}{\hbar}p\cdot y}\psi\left(x+\frac{1}{2}y\right)\overline{\phi\left(x-\frac{1}{2}y\right)}dxdy,$$

we get, using Fubini's theorem,

$$\begin{aligned}\int_{\mathbb{R}^n} W\psi(z)dx &= \left(\frac{1}{2\pi\hbar}\right)^n \int_{\mathbb{R}^{2n}} e^{-\frac{i}{\hbar}p\cdot x}\psi(x')\overline{e^{-\frac{i}{\hbar}p\cdot x}\phi(x)}dxdy \\ &= \left(\frac{1}{2\pi\hbar}\right)^n \int_{\mathbb{R}^n} e^{-\frac{i}{\hbar}p\cdot x}\psi(x)dx\overline{\int_{\mathbb{R}^n} e^{-\frac{i}{\hbar}p\cdot y}\phi(y)dy} \\ &= F\psi(p)\overline{F\phi(p)}\end{aligned}$$

which is formula (3.31). The two formulas (3.29) trivially follow taking $\psi = \phi$. □

Notice that, by integrating with respect to x (*resp.* p) the first (*resp.* second) formula (3.29), we recover the relationship

$$\int_{\mathbb{R}^{2n}} W\psi(z)dz = \|\psi\|_{L^2}.$$

In the quasiprobability distribution, one treats the Wigner transform as if it were a genuine phase-space probability distribution; for instance, given a "quantum observable" $\widehat{A}$ (i. e., a self-adjoint bounded operator) viewed as the quantization of a function a ("classical observable"), one defines the average value (or mean) of $\widehat{A}$ by the formula

$$\langle\widehat{A}\rangle_\psi = \int_{\mathbb{R}^{2n}} a(z)W\psi(z)dz.$$

We will come back to this interpretation in a more rigorous way in the forthcoming chapters.

3.6 Comments and references

The Wigner transform $W\psi$ was introduced by Eugene Wigner in his celebrated 1932 paper [75]. It is unclear what led Wigner (who won the Nobel Prize in physics in 1963) to "guess" his eponymous transform. There are (unconfirmed) speculations that he might have discussed the topic with the physicist Leo Szilard. The importance of the Wigner transform and the associated objects and variants cannot be overestimated. It is at the origin of phase-space quantum mechanics and its formal variant, time-frequency analysis.

The reader interested in a detailed exposition of the properties of the Wigner transform and various developments is invited to consult our monograph [32]. For a slightly different point of view and with different priorities, see the classical treatise by G. Folland [28] which has been seminal to many developments in time-frequency analysis.

Hudson's theorem on the positivity of the Wigner transform was first proven in [47].

4 Gaussians and hermite functions

4.1 The Wigner transform of a gaussian

We will use the following classical formula giving the Fourier transform of a Gaussian:

$$\phi(x) = e^{-\frac{1}{2\hbar}Qx^2}$$

where Q is a symmetric complex $2n \times 2n$ matrix such that $\operatorname{Re} Q > 0$. This condition ensures us that $\phi \in \mathcal{S}(\mathbb{R}^n)$ and hence also $F\phi_M \in \mathcal{S}(\mathbb{R}^n)$:

$$F\phi(x) = (\det Q)^{-1/2}\phi_{Q^{-1}}(x) \tag{4.1}$$

where

$$(\det Q)^{1/2} = \lambda_1^{1/2} \cdots \lambda_m^{1/2},$$

the numbers $\lambda_1^{1/2}, \ldots, \lambda_n^{1/2}$ being the square roots with positive real part of the eigenvalues $\lambda_1, \ldots, \lambda_m$ of Q.

4.1.1 The cross-Wigner transform of a pair of Gaussians

We will henceforth denote by ψ_M the Gaussian function

$$\psi_M = \left(\frac{1}{\pi\hbar}\right)^{n/4} (\operatorname{Re} M)^{1/4}\phi_{M},$$

that is, explicitly:

$$\psi_M(x) = \left(\frac{1}{\pi\hbar}\right)^{n/4} (\det X)^{1/4} e^{-\frac{1}{2\hbar}Mx^2} \tag{4.2}$$

where we have set $M = X + iY$, $X = \operatorname{Re} M > 0$. The prefactor in (4.2) is chosen so that ψ_M is L^2-normalized: $\|\psi_M\|_{L^2} = 1$.

Proposition 49. *Let ψ_M and $\psi_{M'}$ be two Gaussian functions of the type above. We have*

$$W(\psi_M, \psi_{M'})(z) = \left(\frac{1}{\pi\hbar}\right)^{n} C_{M,M'} e^{-\frac{1}{\hbar}Fz\cdot z} \tag{4.3}$$

where $C_{M,M'}$ is a constant:

$$C_{M,M'} = (\det XX')^{1/4} \det\left[\frac{1}{2}(M + \overline{M'})\right]^{-1/2} \tag{4.4}$$

and F is the complex matrix given by

$$F = \begin{pmatrix} 2\overline{M'}(M + \overline{M'})^{-1}M & -i(M - \overline{M'})(M + \overline{M'})^{-1} \\ -i(M + \overline{M'})^{-1}(M - \overline{M'}) & 2(M + \overline{M'})^{-1} \end{pmatrix}. \tag{4.5}$$

https://doi.org/10.1515/9783110722772-004

Proof. We have

$$W(\psi_M,\psi_{M'})(z) = C(X,X')\int_{\mathbb{R}^n} e^{-\frac{i}{\hbar}py}e^{-\frac{1}{2\hbar}\Phi(x,y)}dy$$

with

$$C(X,X') = 2^{-n}\left(\frac{1}{\pi\hbar}\right)^{2n}(\det XX')^{1/4}$$

$$\Phi(x,y) = M\left(x+\frac{1}{2}y\right)^2 + \overline{M'}\left(x-\frac{1}{2}y\right)^2.$$

Let us calculate the integral

$$I(z) = \int_{\mathbb{R}^n} e^{-\frac{i}{\hbar}py}e^{-\frac{1}{2\hbar}\Phi(x,y)}dy.$$

We have

$$\Phi(x,y) = (M+\overline{M'})x^2 + \frac{1}{4}(M+\overline{M'})y^2 + (M-\overline{M'})x\cdot y,$$

and hence,

$$I(z) = e^{-\frac{1}{2\hbar}(M+\overline{M'})x^2}\int_{\mathbb{R}^n} e^{-\frac{i}{\hbar}[p-\frac{i}{2}(M-\overline{M'})x]\cdot y}e^{-\frac{1}{8\hbar}(M+\overline{M'})y^2}dy.$$

Using the Fourier transformation formula (4.1), we get

$$I(z) = (2\pi\hbar)^{n/2}\det\left[\frac{1}{4}(M+\overline{M'})\right]^{-1/2}$$
$$\times\exp\left(-\frac{1}{2\hbar}\left[(M+\overline{M'})x^2 + 4(M+\overline{M'})^{-1}\left(p-\frac{i}{2}(M-\overline{M'})x\right)^2\right]\right).$$

A straightforward calculation shows that

$$\frac{1}{2}(M+\overline{M'})x^2 + 4(M+\overline{M'})^{-1}\left(p-\frac{i}{2}(M-\overline{M'})x\right)^2 = Fz\cdot z$$

where F is the matrix

$$\begin{pmatrix} K & -i(M-\overline{M'})(M+\overline{M'})^{-1} \\ -i(M+\overline{M'})^{-1}(M-\overline{M'}) & 2(M+\overline{M'})^{-1} \end{pmatrix} \tag{4.6}$$

where the left upper block is

$$K = \frac{1}{2}[M+\overline{M'} - (M-\overline{M'})(M+\overline{M'})^{-1}(M-\overline{M'})].$$

Using the identity

$$M + \overline{M'} - (M - \overline{M'})(M + \overline{M'})^{-1}(M - \overline{M'}) = 4\overline{M'}(M + \overline{M'})^{-1}M, \tag{4.7}$$

the matrix (4.6) is given by (4.5). The result follows by collecting the constants and simplifying the obtained expression. □

Suppose that $M = M'$. Then $M + \overline{M'} = 2X$, $M - \overline{M'} = 2Y$, hence (4.5) becomes $F = G$ with

$$G = \begin{pmatrix} X + YX^{-1}Y & YX^{-1} \\ X^{-1}Y & X^{-1} \end{pmatrix}, \tag{4.8}$$

and we have $C_{M,M} = 1$ so formula (4.3) becomes

$$W\psi_M(z) = \left(\frac{1}{\pi\hbar}\right)^n e^{-\frac{1}{\hbar}Gz^2}. \tag{4.9}$$

The following observation is essential for the study of Gaussian quantum states: Besides being obviously symmetric, the matrix G is in addition symplectic: $G \in \mathrm{Sp}(n)$. This follows from the obvious factorization $G = S^T S$ where

$$S = \begin{pmatrix} X^{1/2} & 0 \\ X^{-1/2}Y & X^{-1/2} \end{pmatrix} \in \mathrm{Sp}(n). \tag{4.10}$$

$W\psi_M$ is normalized to unity by making the change of variables $z \longmapsto Sz$ in the integral in formula (4.9).

In particular, when ψ_M is the standard Gaussian

$$\phi_0(x) = \left(\frac{1}{\pi\hbar}\right)^{n/4} e^{-\frac{1}{2\hbar}|x|^2}, \tag{4.11}$$

we immediately get from (4.9)

$$W\phi_0(z) = \left(\frac{1}{\pi\hbar}\right)^n e^{-\frac{1}{\hbar}|z|^2}. \tag{4.12}$$

4.1.2 Fermi's trick

In the case $n = 1$, the standard Gaussian is a solution of the differential equation

$$\frac{1}{2}(-\hbar^2\partial_x^2 + x^2)\phi_0 = \frac{1}{2}\hbar\phi_0 \tag{4.13}$$

(when $\hbar = 1$ this the "Hermite equation"). A natural question to ask is whether the generalized Gaussians (4.2) also are the solutions of a second-order (partial) differential equation. We will use the following elementary result, the proof of which is purely computational, and therefore left to the reader:

Lemma 50. *Let $\psi \in C^2(\mathbb{R}^n)$. Writing $\psi(x) = R(x)e^{i\Phi(x)/\hbar}$ where $R \geq 0$ and Φ are real, that function satisfies the equation $\widehat{H}\psi = 0$ at all points x where $R(x) > 0$ where $\widehat{H}$ is the partial differential operator*

$$\widehat{H} = (-i\hbar\nabla_x - \partial_x\Phi)^2 + \hbar^2\frac{\Delta R}{R}. \tag{4.14}$$

Choosing in particular $\psi = \psi_M$, we have

Proposition 51. *The generalized Gaussian function*

$$\psi_M(x) = \left(\frac{1}{\pi\hbar}\right)^{n/4}(\det X)^{1/4}e^{-\frac{1}{2\hbar}Mx^2}$$

($M = X + iY, X > 0$) satisfies the eigenvalue problem

$$\widehat{H}_M\psi_M = (\hbar \operatorname{Tr} X)\psi_M \tag{4.15}$$

where $\widehat{H}_M$ is the partial differential operator

$$\widehat{H}_M = (-i\hbar\partial_x + Yx)^2 + X^2x\cdot x$$

and $\operatorname{Tr} X > 0$ is the trace of the matrix X.

Proof. Applying Lemma 50 with $\Phi(x) = -\frac{1}{2}Yx \cdot x$ and $R(x) = e^{-\frac{1}{2\hbar}Xx\cdot x}$, we get

$$\partial_x\Phi(x) = -Yx, \quad \frac{\Delta R(x)}{R(x)} = -\frac{1}{\hbar}\operatorname{Tr} X + \frac{1}{\hbar^2}X^2x\cdot x, \tag{4.16}$$

hence the operator $\widehat{H}$ defined by (4.14) is given by

$$\widehat{H} = (-i\hbar\partial_x + Yx)^2 - \hbar \operatorname{Tr} X + X^2x\cdot x\,;$$

formula (4.15) follows since the Weyl symbol of $\widehat{H}$ is the function $H_M - \hbar \operatorname{Tr} X$. □

4.1.3 Cross-ambiguity function of a Gaussian

The calculation of cross-ambiguity functions of Gaussians is straightforward if one uses the algebraic relation (3.22) in Proposition 42:

$$A(\psi,\phi)(z) = 2^{-n}W(\psi,\phi^\vee)\left(\frac{1}{2}z\right)$$

where $\phi^\vee(x) = \phi(-x)$.

Proposition 52. *Let ψ_M and $\psi_{M'}$ be generalized Gaussians (4.2). We have*

$$A(\psi_M, \psi_{M'})(z) = \left(\frac{1}{2\pi\hbar}\right)^n C_{M,M'} e^{-\frac{1}{4\hbar}Fz^2} \tag{4.17}$$

where the constant $C_{M,M'}$ and the matrix F are again given by (4.4) and (4.5). In particular,

$$A\psi_M(z) = \left(\frac{1}{2\pi\hbar}\right)^n e^{-\frac{1}{4\hbar}Gz^2}$$

where the symplectic matrix G is given by (4.8).

Proof. It immediately follows from formula (3.22) in Proposition 42. □

Let for instance $\phi_0(x) = (\pi\hbar)^{-n/4} e^{-|x|^2/2\hbar}$ be the standard Gaussian. Then:

$$A\phi_0(z) = (2\pi\hbar)^{-n} e^{-\frac{1}{4\hbar}|z|^2}.$$

4.2 The case of Hermite functions

4.2.1 Hermite functions

The following formulas are to be found widespread in the literature (see the subsequent comments and references section). The N-th Hermite polynomial h_N is defined by “Rodrigue’s formula”:

$$h_N(x) = (-1)^N e^{x^2}\left(\frac{d^N}{dx^N} e^{-x^2}\right). \tag{4.18}$$

These functions satisfy Hermite’s differential equation

$$\frac{d^2}{dx^2}h_N - 2x\frac{d}{dx}h_N + 2Nh_N = 0 \tag{4.19}$$

and the Hermite recurrence relationships

$$\frac{d}{dx}h_N = 2Nh_{N-1}, \quad h_{N+1} = 2xh_N - 2Nh_{N-1}, \tag{4.20}$$

which allow them to be easily calculated stepwise.

Definition 53. The N-th Hermite function is given by

$$H_N(x) = \sqrt{\frac{1}{2^N N!}}\left(\frac{1}{\pi}\right)^{1/4} e^{-x^2/2} h_N(x). \tag{4.21}$$

Setting $N = 0$, we have $h_0(x) = 1$, hence

$$H_0(x) = \pi^{-1/4} e^{-x^2/2}$$

is the standard Gaussian corresponding to the choice $\hbar = 1$.

The Hermite functions satisfy the parity relationship

$$H_N(-x) = (-1)^N H_N(x), \tag{4.22}$$

and they are the eigenfunctions of the "Hermite operator" $\widehat{H} = \frac{1}{2}(-\partial_x^2 + x^2)$:

$$\frac{1}{2}(-\partial_x^2 + x^2)\phi_N(x) = \left(N + \frac{1}{2}\right)\phi_N(x) \tag{4.23}$$

for $N = 0, 1, 2, \ldots$

In quantum mechanics it is customary to use a rescaled variant of the Hermite functions. Consider the operator

$$\widehat{H} = \frac{1}{2m}(-\hbar^2\partial_x^2 + m^2\omega^2 x^2);$$

it is the quantization of the Hamiltonian function

$$H = \frac{1}{2m}(p^2 + m^2\omega^2 x^2) \tag{4.24}$$

which represents a one-dimensional anisotropic harmonic oscillator with mass m and frequency ω. The operator $\widehat{H}$ has the eigenvalues $\lambda_N = (N + \frac{1}{2})\hbar\omega$, and its eigenfunctions Φ_N are given by

$$\Phi_N(x) = \sqrt{\alpha} H_N(\alpha x), \quad \alpha = \sqrt{\frac{m\omega}{\hbar}},$$

that is, in view of (4.21),

$$\Phi_N(x) = \sqrt{\frac{1}{2^N N!}} \left(\frac{m\omega}{\pi\hbar}\right)^{1/4} e^{-m\omega x^2/2\hbar} h_N\left(\sqrt{\frac{m\omega}{\hbar}} x\right). \tag{4.25}$$

4.2.2 Laguerre polynomials and functions

Let us recall the basics of the theory of Laguerre functions.

Definition 54. The N-th Laguerre polynomial is given by

$$L_N(x) = e^x \frac{d^N}{dx^N}(x^N e^{-x}). \tag{4.26}$$

More generally, for $k \in \mathbb{R}$, the function $L_N^{(k)} : \mathbb{R} \longrightarrow \mathbb{R}$ defined by

$$L_N^{(k)}(x) = \frac{e^x x^{-k}}{N!} \frac{d^N}{dx^N}(x^{N+k} e^{-x}) \tag{4.27}$$

is called the N-th generalized Laguerre polynomial.

Applying Leibniz's rule for the differentiation of a product, we have the explicit expressions

$$L_N(x) = \sum_{j=0}^{N} \binom{N}{j} \frac{(-x)^j}{j!}$$

$$L_N^{(k)}(x) = \sum_{j=0}^{N} \frac{1}{j!} \binom{N+k}{N-j} (-x)^j.$$

The Laguerre polynomials satisfy the Laguerre equation

$$\frac{d^2}{dx^2} L_N^{(k)} + (k+1-x) \frac{d}{dx} L_N^{(k)} + N L_N^{(k)} = 0, \tag{4.28}$$

and we have the following orthogonality relationships:

$$\int_0^\infty x^k e^{-x} L_M^{(k)}(x) L_N^{(k)}(x) dx = \frac{\Gamma(N+k+1)}{N!} \delta_{M,N}. \tag{4.29}$$

Taking $k = 0$, we have in particular:

$$\int_0^\infty e^{-x} L_M(x) L_N(x) dx = \delta_{M,N}. \tag{4.30}$$

The following relationship between Hermite and Laguerre functions is very useful: If $M \leq N$, then

$$\int_{-\infty}^{\infty} e^{-x^2} h_M(x+u) h_N(x+v) dx = 2^n \sqrt{\pi} M! v^{N-M} L_M^{(N-M)}(-2uv)\,; \tag{4.31}$$

choosing $M = N$, we get in particular

$$\int_{-\infty}^{\infty} e^{-x^2} h_N(x+u) h_N(x+v) dx = 2^n \sqrt{\pi} N! L_N(-2uv). \tag{4.32}$$

4.2.3 The cross-Wigner transform of Hermite functions

Proposition 55. *The Wigner transform of the N-th Hermite function Φ_N is given by*

$$W\Phi_N(z) = \frac{(-1)^N}{\pi \hbar} e^{-2H(z)/\hbar\omega} L_N\left(\frac{4}{\hbar\omega} H(z)\right) \tag{4.33}$$

where $z = (x, p)$ and L_N is the N-th simple Laguerre polynomial, H being defined by (4.24).

Proof. We have, by definition of the Wigner transform,

$$W\Phi_N(z) = \frac{1}{2\pi\hbar}\int_{-\infty}^{\infty} e^{-\frac{i}{\hbar}py}\Phi_N\left(x+\frac{1}{2}y\right)\Phi_N\left(x-\frac{1}{2}y\right)dy$$

that is, in view of (4.25)

$$W\Phi_N(z) = \frac{1}{2}C_N\int_{-\infty}^{\infty} e^{-\frac{i}{\hbar}py}e^{-\alpha^2(x^2+\frac{1}{4}y^2)}h_N\left(\alpha\left(x+\frac{1}{2}y\right)\right)h_N\left(\alpha\left(x-\frac{1}{2}y\right)\right)dy$$

$$= C_N e^{-\alpha^2x^2}\int_{-\infty}^{\infty} e^{-\frac{i}{2\hbar}py}e^{-\alpha^2y^2}h_N(\alpha(x+y))h_N(\alpha(x-y))dy$$

where C_N and α are constants:

$$C_N = \frac{1}{\pi\hbar}\frac{1}{2^N N!}\left(\frac{m\omega}{\pi\hbar}\right)^{1/2}, \quad \alpha = \sqrt{\frac{m\omega}{\hbar}}.$$

Completing the squares, we have

$$\alpha^2y^2 + \frac{i}{2\hbar}py = \alpha^2\left(y - \frac{ip}{\alpha^2\hbar}\right)^2 + \frac{p^2}{\alpha^2\hbar^2},$$

and hence

$$W\Phi_N(z) = C_N e^{-\alpha^2x^2}e^{-\frac{p^2}{\alpha^2\hbar^2}}\int_{-\infty}^{\infty} e^{-\alpha^2(y-\frac{ip}{\alpha^2\hbar})^2}h_N(\alpha(x+y))h_N(\alpha(x-y))dy.$$

Setting

$$u = \alpha(y - ip/\alpha^2\hbar) \quad \text{and} \quad \beta = ip/\alpha\hbar,$$

we get, after simplification and using the parity relationship (4.22),

$$W\Phi_N(z) = K_N(z)\int_{-\infty}^{\infty} e^{-u^2}h_N(u+\alpha x+\beta)h_N(u-\alpha x+\beta)du \tag{4.34}$$

where

$$K_N(z) = C_N\frac{(-1)^N}{\alpha}e^{-(\alpha^2x^2-\beta^2)} = \frac{(-1)^N}{\pi\hbar}\frac{\pi^{-1/2}}{2^N N!}e^{-\frac{2}{\hbar\omega}H(z)}.$$

Using formula (4.32), one finds that

$$\int_{-\infty}^{\infty} e^{-u^2} h_N(u + \alpha x + \beta) h_N(u - \alpha x + \beta) du = 2^N \sqrt{\pi} N! L_N\left(\frac{4}{\hbar\omega} H(z)\right)$$

where L_N is the Laguerre polynomial (4.26), hence, after some simplifications,

$$W\Phi_N(z) = \frac{(-1)^N}{\pi\hbar} e^{-\frac{2}{\hbar\omega}H(z)} L_N\left(\frac{4}{\hbar\omega}H(z)\right)$$

which is formula (4.33). □

If we choose $m = \omega = 1$, then $H = \frac{1}{2}(p^2 + x^2)$, and (4.33) becomes

$$W\Phi_N(x,p) = \frac{(-1)^N}{\pi\hbar} e^{-\frac{1}{\hbar}|z|^2} L_N\left(\frac{2}{\hbar}|z|^2\right). \tag{4.35}$$

Taking in addition $\hbar = 1$ we get the standard Hermite function H_N and

$$WH_N(x,p) = \frac{(-1)^N}{\pi} e^{-|z|^2} L_N(2|z|^2), \tag{4.36}$$

while the choice $\hbar = 1/2\pi$ leads to

$$W\Phi_N(x,p) = 2(-1)^N e^{-2\pi|z|^2} L_N(4\pi|z|^2). \tag{4.37}$$

The main result of this section is:

Proposition 56. *Let $M \leq N$. The cross-Wigner transform of Φ_M and Φ_N is given by*

$$W(\Phi_M, \Phi_N)(z) = C_{M,N}\left(x - \frac{ip}{m\omega}\right)^{N-M} e^{-\frac{2}{\hbar\omega}H(z)} L_M^{(N-M)}\left(\frac{4}{\hbar\omega}H(z)\right) \tag{4.38}$$

where the constant factor $C_{M,N}$ is given by

$$C_{M,N} = \frac{(-1)^M}{\pi\hbar}\sqrt{\frac{2^N M!}{2^M N!}}\left(\frac{m\omega}{\hbar}\right)^{(N-M)/2}.$$

Proof. In the course of the proof of Proposition 55, we showed that

$$W\Phi_N(z) = K_N(z) \int_{-\infty}^{\infty} e^{-u^2} h_N(u + \alpha x + \beta) h_N(u - \alpha x + \beta) du \tag{4.39}$$

where

$$K_N(z) = \frac{(-1)^N}{\pi\hbar} \frac{\pi^{-1/2}}{2^N N!} e^{-\frac{2}{\hbar\omega}H(z)}.$$

In the present case, a similar argument leads to the formula

$$W(\Phi_M, \Phi_N)(z) = K_{M,N}(z) I_{M,N}(z)$$

where the factor $K_{M,N}(z)$ is given by

$$K_{M,N}(z) = \frac{(-1)^N}{\pi\hbar}\sqrt{\frac{1}{2^{M+N}M!N!}}\pi^{-1/2}e^{-\frac{2}{\hbar\omega}H(z)}$$

and

$$I_{M,N}(z) = \int_{-\infty}^{\infty} e^{-u^2} h_M(u + \alpha x + \beta) h_N(u - \alpha x + \beta)du.$$

In view of formula (4.32), we have

$$I_{M,N}(z) = 2^N \sqrt{\pi} M!(-1)^{N-M}(\alpha x - \beta)^{N-M} L_M^{(N-M)}(2(\alpha^2 x^2 - \beta^2));$$

using the relationships

$$\alpha x - \beta = \sqrt{\frac{m\omega}{\hbar}}\left(x - \frac{ip}{m\omega}\right)$$

$$2(\alpha^2 x^2 - \beta^2) = \frac{4}{\hbar\omega}H(z),$$

the function $I_{M,N}(z)$ is given by

$$I_{M,N}(z) = 2^N M!(-1)^{N-M}\left(\frac{m\omega}{\hbar}\right)^{(N-M)/2}\left(x - \frac{ip}{m\omega}\right)^{N-M} L_M^{(N-M)}\left(\frac{4}{\hbar\omega}H(z)\right),$$

hence formula (4.38) after a few simplifications. □

If we choose $m = \omega = 1$ and $\hbar = 1/2\pi$, formula (4.38) can be written

$$W(\Phi_M, \Phi_N)(z) = C_{M,N}\overline{\zeta}^{N-M} e^{-2\pi|\zeta|^2} L_M^{(N-M)}(4\pi|\zeta|^2) \tag{4.40}$$

where ζ is the complex variable $\zeta = x + ip$ and the constant $C_{M,N}$ is given by

$$C_{M,N} = 2^{N-M+1}(-1)^M \pi^{(N-M)/2}\sqrt{\frac{M!}{N!}}.$$

The following formulas for the cross-ambiguity function of (Φ_M, Φ_N) easily follow from the preceding result:

Corollary 57. *Let $M \leq N$. We have*

$$A(\Phi_M, \Phi_N)(z) = D_{M,N}\left(x - \frac{ip}{m\omega}\right)^{N-M} e^{-\frac{1}{2\hbar\omega}H(z)} L_M^{(N-M)}\left(\frac{1}{\hbar\omega}H(z)\right) \tag{4.41}$$

where

$$D_{M,N} = \frac{(-1)^{M+N}}{2\pi\hbar} 2^{N-M} \sqrt{\frac{2^N M!}{2^M N!}} \left(\frac{m\omega}{\hbar}\right)^{(N-M)/2}.$$

In particular,

$$A\Phi_N(z) = \frac{1}{2\pi\hbar} e^{-\frac{1}{2\hbar\omega}H(z)} L_N\left(\frac{1}{\hbar\omega}H(z)\right). \tag{4.42}$$

Proof. The relationship (3.22) in Proposition 42 becomes here

$$A(\psi, \phi)(z) = \frac{1}{2} W(\psi, \phi^\vee)\left(\frac{1}{2}z\right). \tag{4.43}$$

Using the parity formula (4.22), we have $\Phi_N^\vee(x) = (-1)^N \Phi_N(x)$, hence

$$A(\Phi_M, \Phi_N)(z) = \frac{1}{2}(-1)^N W(\Phi_M, \Phi_N)\left(\frac{1}{2}z\right).$$ □

4.3 Comments and references

Hudson showed in [47] that the Wigner transform of a function on ℝ is positive if an only that function is a generalized Gaussian. Janssen generalized in [49] Hudson's result to the multidimensional case under more general hypotheses. See Folland [28], Chap. 1, Gröchenig [40], Soto and Claverie [67], Toft [70] for independent proofs. We refer to the monumental treatise [39] by Gradshteyn and Ryzhik for material on Hermite and Laguerre functions. The cross-Wigner transform of Hermite functions has been studied by Agorram et al. [3]; also see our monograph [32] which we have been following in our presentation. One can find variants of the proof of the formula for the Wigner transform of Hermite functions in many texts.

5 The Weyl transform

5.1 Definitions of the Weyl transform

The Weyl transform associates to every symbol $a \in \mathcal{S}'(\mathbb{R}^{2n})$ an operator on $\mathcal{S}(\mathbb{R}^n)$. The properties of the Weyl transform make it into a possible quantization procedure.

5.1.1 First definition

In Proposition 179, we expressed the relationship between a density operator $\widehat{\rho}$ and its Wigner distribution ρ using the harmonic decomposition

$$\widehat{\rho}\psi = \int_{\mathbb{R}^{2n}} \rho_\sigma(z)\widehat{D}(z)\psi dz \tag{5.1}$$

where $\psi \in L^2(\mathbb{R}^n)$, $\widehat{D}(z)$ is the Heisenberg displacement operator, and $\rho_\sigma = F_\sigma \rho$ is the symplectic Fourier transform of ρ. Equivalently, if one uses the Grossmann–Royer reflection operator,

$$\widehat{\rho}\psi = 2^n \int_{\mathbb{R}^{2n}} \rho(z)\widehat{R}(z)\psi dz. \tag{5.2}$$

This suggests the following definition.

Definition 58. Let $a \in \mathcal{S}'(\mathbb{R}^{2n})$. The Weyl operator $\mathrm{Op_W}(a)$ with symbol a is the mapping $\mathcal{S}(\mathbb{R}^n) \longrightarrow \mathcal{S}'(\mathbb{R}^n)$ defined by

$$\mathrm{Op_W}(a)\psi = \left(\frac{1}{2\pi\hbar}\right)^n \int_{\mathbb{R}^{2n}} a_\sigma(z)\widehat{D}(z)\psi dz, \tag{5.3}$$

and the mapping $a \longmapsto \mathrm{Op_W}(a)$ is called the “Weyl transform”; formula (5.3) is equivalent to

$$\mathrm{Op_W}(a)\psi = \left(\frac{1}{\pi\hbar}\right)^n \int_{\mathbb{R}^{2n}} a(z)\widehat{R}(z)\psi dz. \tag{5.4}$$

[Some authors use the notation $a^w(x, D)$ instead of $\mathrm{Op_W}(a)$.]

The proof of the equivalence of the definitions (5.3) and (5.4) is *mutatis mutandis* similar to that of Proposition 179 using the symplectic Fourier transform. These formulas should be understood in the distributional sense: for every $\phi \in \mathcal{S}(\mathbb{R}^n)$, we have

$$\langle \mathrm{Op_W}(a)\psi, \phi\rangle = \left(\frac{1}{2\pi\hbar}\right)^n \langle a_\sigma, \langle \widehat{D}(\cdot)\psi, \phi\rangle\rangle$$

https://doi.org/10.1515/9783110722772-005

$$\langle \mathrm{Op_W}(a)\psi, \phi\rangle = \left(\frac{1}{\pi\hbar}\right)^n \langle a, \langle \widehat{R}(\cdot)\psi, \phi\rangle\rangle$$

where $\langle \widehat{D}(\cdot)\psi, \phi\rangle \in \mathcal{S}(\mathbb{R}^{2n})$ and $\langle \widehat{R}(\cdot)\psi, \phi\rangle \in \mathcal{S}(\mathbb{R}^{2n})$.

Remark 59. The density operator $\widehat{\rho}$ defined by (5.1) is thus a Weyl operator with symbol $(2\pi\hbar)^n\rho$.

Admittedly, this definition is a little vague since one does not immediately see how to interpret the integral in (9.37) except in a weak sense, nor is it clear what the domain (and the range) of $\mathrm{Op_W}(a)$ could be. Also, the presence of the prefactor $(2\pi\hbar)^{-n}$, which does not appear in (5.1), has to be motivated. So, a few words of explanation are necessary.

Proposition 60. (i) *The Weyl transform* $\mathrm{Op_W}(a)$ *is a well-defined operator* $\mathcal{S}(\mathbb{R}^n) \longrightarrow \mathcal{S}'(\mathbb{R}^n)$ *for all* $a \in \mathcal{S}'(\mathbb{R}^{2n})$ *and* (ii) *Let* $b \in \mathcal{S}'(\mathbb{R}^n)$*; then*

$$\mathrm{Op_W}(b \otimes 1)\psi(x) = b(x)\psi(x),$$

and, in particular, $\mathrm{Op_W}(1)$ *is the identity operator* I_{d}.

Proof. (i) Let $\psi \in \mathcal{S}(\mathbb{R}^n)$. For every $\phi \in \mathcal{S}(\mathbb{R}^n)$, the distributional bracket $\langle \widehat{D}(z)\psi, \phi\rangle$ is equal to $(\pi\hbar)^n\overline{A(\psi,\phi)}(z)$ (A the cross-ambiguity function) and hence, since $A(\psi,\phi) \in \mathcal{S}(\mathbb{R}^{2n})$ (see (3.20))

$$\begin{aligned}\langle \mathrm{Op_W}(a)\psi, \phi\rangle &= 2^{-n}\int_{\mathbb{R}^{2n}} \rho_\sigma(z)\overline{A(\psi,\phi)}(z)dz\\ &= 2^{-n}\langle \rho_\sigma, \overline{A(\psi,\phi)}\rangle\end{aligned}$$

hence our claim since $a \in \mathcal{S}'(\mathbb{R}^{2n})$ if and only if $a_\sigma \in \mathcal{S}'(\mathbb{R}^{2n})$. (ii) It is sufficient to give the proof for $b \in \mathcal{S}(\mathbb{R}^n)$. We have, using the equality $F_n 1 = (2\pi\hbar)^{n/2}\delta$ (F_n the n-dimensional Fourier transform),

$$\begin{aligned}\mathrm{Op_W}(b\otimes 1)\psi(x') &= \left(\frac{1}{2\pi\hbar}\right)^{n/2}\int_{\mathbb{R}^{2n}} \widehat{b}(p)\delta(x)\widehat{D}(z)\psi(x')dz\\ &= \left(\frac{1}{2\pi\hbar}\right)^{n/2}\int_{\mathbb{R}^n} e^{\frac{i}{\hbar}px'}\widehat{b}(p)\psi(x')dp\\ &= b(x')\psi(x').\end{aligned}$$

In the case where the symbol is the Dirac distribution δ on $\mathbb{R}^{2n}$, we have $a_\sigma = (2\pi\hbar)^{-n}$, and hence

$$\mathrm{Op_W}(\delta)\psi(x) = \left(\frac{1}{2\pi\hbar}\right)^{2n}\int_{\mathbb{R}^n}\left(\int_{\mathbb{R}^n} e^{\frac{i}{\hbar}(p_0\cdot x - \frac{1}{2}p_0\cdot x_0)}dp_0\right)\psi(x-x_0)dx_0$$

$$= \left(\frac{1}{2\pi\hbar}\right)^n \int_{\mathbb{R}^n} \delta\left(x - \frac{1}{2}x_0\right)\psi(x - x_0)dx_0$$

$$= \left(\frac{1}{\pi\hbar}\right)^n \psi(-x)$$

so that $\mathrm{Op_W}(\delta)$ is, up to the factor $(\pi\hbar)^{-n}$, the reflection operator $\widehat{R} = \widehat{R}(0)$. □

Another consequence of this definition is:

Proposition 61. *The operator with Weyl symbol $z \mapsto e^{-\frac{i}{\hbar}\sigma(z,z_0)}$ is the Heisenberg displacement operator $\widehat{D}(z_0)$.*

Proof. Let us write $a_0(z) = e^{-\frac{i}{\hbar}\sigma(z,z_0)}$, and let $\widehat{A}_0 = \mathrm{Op_W}(a_0)$. We have

$$\widehat{A}_0\psi(x) = \left(\frac{1}{\pi\hbar}\right)^n \int_{\mathbb{R}^{2n}} e^{-\frac{i}{\hbar}\sigma(z,z')} e^{\frac{2i}{\hbar}p_0\cdot(x-x_0)}\psi(2x_0 - x)dp_0dx_0$$

$$= \left(\frac{1}{\pi\hbar}\right)^n \int_{\mathbb{R}^{2n}} e^{\frac{i}{\hbar}p'\cdot x_0} e^{\frac{i}{\hbar}p_0\cdot(x'+2x_0-2x)}\psi(2x_0 - x)dp_0dx_0$$

$$= \left(\frac{1}{\pi\hbar}\right)^n \int_{\mathbb{R}^n} e^{\frac{i}{\hbar}p'\cdot x_0}\left[\int_{\mathbb{R}^n} e^{\frac{i}{\hbar}p_0\cdot(x'+2x_0-2x)}dp_0\right]\psi(2x_0 - x)dx_0$$

$$= 2^n \int_{\mathbb{R}^n} e^{\frac{i}{\hbar}p'\cdot x_0}\delta(x' + 2x_0 - 2x)\psi(2x_0 - x)dx_0.$$

Setting $y = 2x_0$, we get

$$\widehat{A}_0\psi(x) = e^{\frac{i}{\hbar}(p'\cdot x - \frac{1}{2}p'\cdot x')}\psi(x - x') = \widehat{D}(z_0)\psi(x).$$ □

5.1.2 Definition using the Wigner transform

It is a variant of the previous definition.

Definition 62. Let $(\psi,\phi) \in \mathcal{S}(\mathbb{R}^n) \times \mathcal{S}(\mathbb{R}^n)$, and $a \in \mathcal{S}'(\mathbb{R}^{2n})$. Then the Weyl operator $\mathrm{Op_W}(a)$ is the only operator $\mathcal{S}(\mathbb{R}^n) \longrightarrow \mathcal{S}'(\mathbb{R}^n)$ such that

$$\langle \mathrm{Op_W}(a)\psi, \overline{\phi}\rangle = \langle a, W(\psi,\phi)\rangle. \tag{5.5}$$

An operator satisfying (5.5) (if it exits) is unique: If $\widehat{A} : \mathcal{S}(\mathbb{R}^n) \longrightarrow \mathcal{S}'(\mathbb{R}^n)$ is another such operator, then

$$\langle (\widehat{A} - \mathrm{Op_W}(a))\psi, \overline{\phi}\rangle = 0$$

for all ϕ hence $(\widehat{A} - \mathrm{Op_W}(a))\psi = 0$ for all ψ which is only possible if $\widehat{A} = \mathrm{Op_W}(a)$. Let us show that such an operator $\widehat{A}$ exists; for this it suffices to check that

$$\widehat{A}\psi = \left(\frac{1}{\pi\hbar}\right)^n \int_{\mathbb{R}^{2n}} a(z)\widehat{R}(z)\psi dz$$

for $\psi \in \mathcal{S}(\mathbb{R}^n)$. Letting $\phi \in \mathcal{S}(\mathbb{R}^n)$, we have using definition (3.1) of the cross-Wigner transform,

$$\begin{aligned}\langle \widehat{A}\psi, \overline{\phi}\rangle &= \left(\frac{1}{\pi\hbar}\right)^n \int_{\mathbb{R}^{2n}} a(z)\langle \widehat{R}(z)\psi, \overline{\phi}\rangle dz \\ &= \int_{\mathbb{R}^{2n}} a(z)W(\psi,\phi)(z)dz\end{aligned}$$

[the integrals being viewed as distributional brackets on $\mathbb{R}^{2n}$] which proves our claim.

An immediate consequence of this definition is:

Proposition 63. *If $a \in L^2(\mathbb{R}^{2n})$, then $\mathrm{Op_W}(a): L^2(\mathbb{R}^n) \longrightarrow L^2(\mathbb{R}^n)$ continuously.*

Proof. Definition (5.5) implies that

$$(\mathrm{Op_W}(a)\psi|\phi)_{L^2} = (a|\overline{W(\psi,\phi)})_{L^2(\mathbb{R}^{2n})} \tag{5.6}$$

for all $(\psi,\phi) \in L^2(\mathbb{R}^n) \times L^2(\mathbb{R}^n)$, since we have in this case $W(\psi,\phi) \in L^2(\mathbb{R}^{2n})$ (Proposition 38). In view of the Cauchy–Schwarz inequality together with the Moyal identity (3.12), we have

$$\begin{aligned}|(a|\overline{W(\psi,\phi)})_{L^2(\mathbb{R}^{2n})}|^2 &\le \|a\|_{L^2(\mathbb{R}^{2n})}\|W(\psi,\phi)\|_{L^2(\mathbb{R}^{2n})} \\ &= \left(\frac{1}{2\pi\hbar}\right)^{n/2}\|a\|_{L^2(\mathbb{R}^{2n})}\|\psi\|_{L^2|}\|\phi\|_{L^2}\end{aligned}$$

that is, taking $\psi = \phi$ in (5.6),

$$|(\mathrm{Op_W}(a)\psi|\psi)_{L^2}| \le \left(\frac{1}{2\pi\hbar}\right)^{n/2}\|a\|_{L^2(\mathbb{R}^{2n})}\|\psi\|^2_{L^2|},$$

hence the result. Here is another application: One consequence of Moyal's identity is that, if (ψ_j) is an orthonormal set of vectors in $L^2(\mathbb{R}^n)$, then the functions $W(\psi_j,\psi_k)$ are orthogonal as well. In fact, more precisely: □

Proposition 64. *Let $(\psi_j)_j$ and $(\phi_j)_j$ be orthonormal bases of $L^2(\mathbb{R}^n)$. The family of functions $(\Phi_{j,k})_{j,k}$ with $\Phi_{j,k} = (2\pi\hbar)^{n/2}W(\psi_j,\phi_{jk})$ is an orthonormal basis of $L^2(\mathbb{R}^{2n})$.*

Proof. In view of the Moyal identity (3.12) for the cross-Wigner transform, we have

$$(\Phi_{j,k}|\Phi_{j',k'})_{L^2(\mathbb{R}^{2n})} = (\psi_j|\psi_{j'})_{L^2}\overline{(\phi_k|\phi_{k'})_{L^2}},$$

hence the $\Phi_{j,k}$ form an orthonormal system in $L^2(\mathbb{R}^{2n})$. To show that $(\Phi_{j,k})_{j,k}$ is a basis of $L^2(\mathbb{R}^{2n})$, we must show that the vectors $\Phi_{j,k}$ span $L^2(\mathbb{R}^{2n})$. For this, it is sufficient to show that, if $a \in L^2(\mathbb{R}^{2n})$ is orthogonal to each $\Phi_{j,k}$, then $a = 0$. Assume that $(a|\Phi_{jk})_{L^2(\mathbb{R}^{2n})} = 0$ for all indices j, k. Since we have (formula (5.6))

$$\begin{aligned}(a|\Phi_{jk})_{L^2(\mathbb{R}^{2n})} &= (2\pi\hbar)^{n/2}(a|W(\psi_j, \phi_k))_{L^2(\mathbb{R}^{2n})} \\ &= (2\pi\hbar)^{n/2}(\operatorname{Op}_{\mathrm{W}}(a)\phi_k|\psi_j)_{L^2(\mathbb{R}^{2n})},\end{aligned}$$

the condition $(a|\Phi_{jk})_{L^2(\mathbb{R}^{2n})} = 0$ for all j, k is equivalent to

$$(\operatorname{Op}_{\mathrm{W}}(a)\phi_k|\psi_j)_{L^2(\mathbb{R}^{2n})} = 0$$

for all indices j, k, and this implies that $\operatorname{Op}_{\mathrm{W}}(a) = 0$ since $(\psi_j)_j$ and $(\phi_k)_k$ are bases of $L^2(\mathbb{R}^n)$. In view of the uniqueness of the Weyl symbol of an operator, this implies in turn that $a = 0$ as we set out to prove. □

5.1.3 Pseudo-differential definition

The pseudo-differential definition of the Weyl transform is the most commonly used in introductory texts obtained by taking literally formula (5.2):

Definition 65. Let $a \in \mathcal{S}(\mathbb{R}^{2n})$. The Weyl operator with symbol a is defined by

$$\operatorname{Op}_{\mathrm{W}}(a)\psi(x) = \left(\frac{1}{2\pi\hbar}\right)^n \int_{\mathbb{R}^n\times\mathbb{R}^n} e^{\frac{i}{\hbar}p\cdot(x-y)} a\left(\frac{1}{2}(x+y), p\right)\psi(y)dpdy \tag{5.7}$$

for $\psi \in \mathcal{S}(\mathbb{R}^n)$.

This definition makes immediately visible that Weyl operators are Fourier integral operators of a particular type. Its main drawback is that it is not immediately obvious from formula (5.7) how to extend the definition to symbols that are not decreasing fast enough to make the integral absolutely convergent. Things become more convincing if one remarks that (5.7) is the integral form of the identity (5.4). In fact, using the explicit expression (2.9) of the Grossmann–Royer reflection operator, we get

$$\begin{aligned}\operatorname{Op}_{\mathrm{W}}(a)\psi(x') &= \left(\frac{1}{\pi\hbar}\right)^n \int_{\mathbb{R}^{2n}} a(z)\widehat{R}(z)\psi(x')dz \\ &= \left(\frac{1}{\pi\hbar}\right)^n \int_{\mathbb{R}^{2n}} a(x,p)e^{\frac{2i}{\hbar}p\cdot(x'-x)}\psi(2x-x')dpdx\,;\end{aligned}$$

setting $y = 2x - x'$, we get

$$\operatorname{Op}_{\mathrm{W}}(a)\psi(x') = \left(\frac{1}{2\pi\hbar}\right)^n \int_{\mathbb{R}^{2n}} a\left(\frac{1}{2}(x'+y), p\right)e^{\frac{i}{\hbar}p\cdot(x'-y)}\psi(y)dpdy,$$

which is (5.7).

There are several ways to reinterpret the integral in (5.7) when it is not absolutely convergent; perhaps the simplest is to use a compactly supported cut-off function $\chi \in C_0^\infty(\mathbb{R}^{3n})$ such that $\chi(0,0,0) = 1$ and to define

$$\mathrm{Op_W}(a)\psi(x) = \lim_{\varepsilon \to 0} I_\varepsilon(x) \tag{5.8}$$

where

$$I_\varepsilon(x) = \left(\frac{1}{2\pi\hbar}\right)^n \int_{\mathbb{R}^{2n}} e^{\frac{i}{\hbar}p\cdot(x-y)} \chi(\varepsilon x, \varepsilon y, \varepsilon p) a\left(\frac{1}{2}(x+y), p\right)\psi(y)dpdy.$$

The integral $I_\varepsilon(x)$ is absolutely convergent provided that, for instance, $\chi a \in L^1(\mathbb{R}^{2n})$. Both definitions (5.7) and (5.8) coincide when $a \in \mathcal{S}(\mathbb{R}^{2n})$ and $\psi \in \mathcal{S}(\mathbb{R}^n)$; in this case, the integrals (which we denote respectively $I(x)$ and $I_\varepsilon(x)$) in (5.7) and (5.8) are absolutely convergent, and we have

$$|I_1(x) - I_\varepsilon(x)| \leq \int_{\mathbb{R}^{2n}} |1 - \chi(\varepsilon x, \varepsilon y, \varepsilon p)| \left| a\left(\frac{1}{2}(x+y), p\right)\right| |\psi(y)| dpdy,$$

hence $\lim_{\varepsilon \to 0} |I_1(x) - I_\varepsilon(x)| = 0$ by Lebesgue's dominated convergence theorem. In particular, the choice of the cut-off function χ is irrelevant.

5.2 Main properties

5.2.1 The distributional kernel of a Weyl operator

Let $\widehat{A} : \mathcal{S}(\mathbb{R}^n) \longrightarrow \mathcal{S}'(\mathbb{R}^n)$ be an operator. If there exists a distribution $K \in \mathcal{S}'(\mathbb{R}^n \times \mathbb{R}^n)$ such that

$$\langle \widehat{A}\psi, \phi \rangle = \langle K, \phi \otimes \psi \rangle$$

for all $\phi, \psi \in \mathcal{S}(\mathbb{R}^n)$, then K is called the (distributional, or Schwartz) *kernel* of $\widehat{A}$. We will also write informally

$$A\psi(x) = \int_{\mathbb{R}^n} K(x,y)\psi(y)dy$$

where the integral is to be understood in the distributional sense. It turns out that, in fact, every *continuous* operator $\widehat{A} : \mathcal{S}(\mathbb{R}^n) \longrightarrow \mathcal{S}'(\mathbb{R}^n)$ has a kernel $K \in \mathcal{S}'(\mathbb{R}^n \times \mathbb{R}^n)$ (this is Schwartz's kernel theorem).

We are going to see that every continuous operator $\mathcal{S}(\mathbb{R}^n) \longrightarrow \mathcal{S}'(\mathbb{R}^n)$ is a Weyl operator $\mathrm{Op_W}(a)$ for some symbol $a \in \mathcal{S}'(\mathbb{R}^{2n})$.

Proposition 66. *Let $a \in \mathcal{S}'(\mathbb{R}^{2n})$.* (i) *The kernel of* $\operatorname{Op_W}(a)$ *is given by*

$$K(x,y) = \left(\frac{1}{2\pi\hbar}\right)^n \int_{\mathbb{R}^n} e^{\frac{i}{\hbar}p\cdot(x-y)} a\left(\frac{1}{2}(x+y), p\right) dp \tag{5.9}$$

interpreted as a Fourier transform in the variable p. (ii) *If conversely $\widehat{A}$ is a continuous operator $\mathcal{S}(\mathbb{R}^n) \longrightarrow \mathcal{S}'(\mathbb{R}^n)$ with kernel K, then $\widehat{A} = \operatorname{Op_W}(a)$ with*

$$a(x,p) = \int_{\mathbb{R}^n} e^{-\frac{i}{\hbar}p\cdot y} K\left(x + \frac{1}{2}y, x - \frac{1}{2}y\right) dy \tag{5.10}$$

interpreted as the inverse Fourier transform in y.

Proof. The result is trivial using definition (5.7) of a Weyl operator and the relationship

$$K\left(x + \frac{1}{2}y, x - \frac{1}{2}y\right) = \left(\frac{1}{2\pi\hbar}\right)^n \int_{\mathbb{R}^n} e^{\frac{i}{\hbar}p\cdot y} a(x,p) dp. \tag{5.11}$$

□

5.2.2 The adjoint and the transpose of a Weyl operator

Assume that $\widehat{A}$ is an operator $\mathcal{S}(\mathbb{R}^n) \longrightarrow \mathcal{S}(\mathbb{R}^n)$. The transposed operator $\widehat{A}^T$ is defined by the formula

$$\langle \widehat{A}\psi, \phi\rangle = \langle \psi, \widehat{A}^T\phi\rangle \tag{5.12}$$

for $(\psi, \phi) \in \mathcal{S}(\mathbb{R}^n) \times \mathcal{S}(\mathbb{R}^n)$ and the adjoint $\widehat{A}^*$ by

$$(\widehat{A}^*\psi|\phi)_{L^2} = (\psi|\widehat{A}\phi)_{L^2}.$$

Proposition 67. (i) *The transpose $\widehat{A}^T$ of $\widehat{A} = \operatorname{Op_W}(a)$ is the Weyl operator $\widehat{A}^T = \operatorname{Op_W}(b)$ where the symbol b is obtained from a by the partial reflection $p \longmapsto -p$: , that is,*

$$b(z) = a(\overline{z}) \quad \text{with } \overline{z} = (x,p). \tag{5.13}$$

(ii) *The adjoint $\widehat{A}^*$ is obtained by complex conjugation of the Weyl symbol:*

$$\widehat{A}^* = \operatorname{Op_W}(\overline{a}). \tag{5.14}$$

Proof. (i) By formula (5.4), we have, taking the transposition formula (2.10) for the reflection operator into account,

$$\langle \widehat{A}\psi, \phi\rangle = \left(\frac{1}{\pi\hbar}\right)^n \int_{\mathbb{R}^{2n}} a(z)\langle \widehat{R}(z)\psi, \phi\rangle dz$$

$$= \left(\frac{1}{\pi\hbar}\right)^n \int_{\mathbb{R}^{2n}} a(z)\langle\psi, \widehat{R}(\overline{z})\phi\rangle dz$$
$$= \left(\frac{1}{\pi\hbar}\right)^n \int_{\mathbb{R}^{2n}} a(\overline{z})\langle\widehat{R}(z)\phi, \psi\rangle d\overline{z},$$

hence, since $d\overline{z} = dz$,

$$\widehat{A}^T\phi = \left(\frac{1}{\pi\hbar}\right)^n \int_{\mathbb{R}^{2n}} a(\overline{z})\widehat{R}(z)\phi dz$$

which proves (5.12). (ii) Formula (5.14) is proven by a similar argument, replacing the distributional brackets with the scalar product on $L^2(\mathbb{R}^n)$. □

5.3 Shubin's symbol classes

For $z \in \mathbb{R}^{2n}$, we set $\langle z\rangle = (1 + |z|^2)^{1/2}$.

Definition 68. Let $m \in \mathbb{R}$ and $\rho \in]0, 1]$. The Shubin symbol class $\Gamma_\rho^m(\mathbb{R}^{2n})$ consists of all complex functions $a \in C^\infty(\mathbb{R}^{2n})$ such that for every $\alpha \in \mathbb{N}^{2n}$ there exists a constant $C_\alpha \geq 0$ with

$$|\partial_z^\alpha a(z)| \leq C_\alpha \langle z\rangle^{m-\rho|\alpha|} \quad \text{for } z \in \mathbb{R}^{2n} \tag{5.15}$$

where We denote by $G_\rho^m(\mathbb{R}^n)$ the vector space of all Weyl operators $\mathrm{Op_W}(a)$ with $a \in \Gamma_\rho^m(\mathbb{R}^{2n})$.

Obviously $\Gamma_\rho^m(\mathbb{R}^{2n})$ is a complex vector space for the usual operations of addition and multiplication by complex numbers. Using the generalized Leibniz rule for the derivatives of a product of functions, one easily checks that:

$$a \in \Gamma_\rho^m(\mathbb{R}^{2n}) \text{ and } b \in \Gamma_\rho^{m'}(\mathbb{R}^{2n}) \Longrightarrow ab \in \Gamma_\rho^{m+m'}(\mathbb{R}^{2n});$$
$$a \in \Gamma_\rho^m(\mathbb{R}^{2n}) \text{ and } \alpha \in \mathbb{N}^{2n} \Longrightarrow \partial_z^\alpha a \in \Gamma_\rho^{m-|\alpha|}(\mathbb{R}^{2n}).$$

One also easily verifies that

$$\Gamma_\rho^{-\infty}(\mathbb{R}^{2n}) = \bigcap_{m\in\mathbb{R}} \Gamma_\rho^m(\mathbb{R}^{2n}) = \mathcal{S}(\mathbb{R}^{2n}).$$

The symbol class $\Gamma_\rho^m(\mathbb{R}^{2n})$ is preserved by linear changes of variables: If $a \in \Gamma_\rho^m(\mathbb{R}^{2n})$ and M is a linear automorphism of $\mathbb{R}^{2n}$, then $a \circ M$ is also in $\Gamma_\rho^m(\mathbb{R}^{2n})$ (the elementary proof is left to the reader).

The elements of $G_\rho^m(\mathbb{R}^n)$ have many interesting regularity properties; for instance:

Proposition 69. *Every Weyl operator* $\operatorname{Op_W}(a) \in G^m_\rho(\mathbb{R}^n)$ *is a continuous operator* $\mathcal{S}(\mathbb{R}^n) \longrightarrow \mathcal{S}(\mathbb{R}^n)$ *extending into a continuous operator* $\mathcal{S}'(\mathbb{R}^n) \longrightarrow \mathcal{S}'(\mathbb{R}^n)$.

Proof. That $\widehat{A} = \operatorname{Op_W}(a)$ extends into a continuous operator $\mathcal{S}'(\mathbb{R}^n) \longrightarrow \mathcal{S}'(\mathbb{R}^n)$ follows from the first statement by duality since the transpose $\widehat{A}^T$ is also in $G^m_\rho(\mathbb{R}^n)$. In fact, the condition $a \in \Gamma^m_\rho(\mathbb{R}^{2n})$ implies that $b \in \Gamma^m_\rho(\mathbb{R}^{2n})$ where $b(z) = a(\bar{z})$ (Proposition 67), hence $\widehat{A}^T : \mathcal{S}(\mathbb{R}^n) \longrightarrow \mathcal{S}(\mathbb{R}^n)$. One can then define the extension of $\widehat{A}$ (also denoted $\widehat{A}$) by the formula $\langle \widehat{A}\psi, \phi\rangle = \langle \psi, \widehat{A}^T\phi\rangle$ for $\psi \in \mathcal{S}'(\mathbb{R}^n)$. The continuity $\mathcal{S}(\mathbb{R}^n) \longrightarrow \mathcal{S}(\mathbb{R}^n)$ is easily verified using the estimates on the semi-norms together with the polynomial boundedness of the derivatives of the symbol $a \in \Gamma^m_\rho(\mathbb{R}^{2n})$. □

5.4 Composition formulas

We now assume that the Weyl operators

$$\widehat{A} = \operatorname{Op_W}(a) = \left(\frac{1}{2\pi\hbar}\right)^n \int_{\mathbb{R}^{2n}} a_\sigma(z)\widehat{D}(z)dz$$

$$\widehat{B} = \operatorname{Op_W}(b) = \left(\frac{1}{2\pi\hbar}\right)^n \int_{\mathbb{R}^{2n}} b_\sigma(z)\widehat{D}(z)dz$$

can be composed. The compose $\widehat{C} = \widehat{A}\widehat{B}$ is, for instance, always defined when

$$\operatorname{Op_W}(a) : \mathcal{S}(\mathbb{R}^n) \longrightarrow \mathcal{S}'(\mathbb{R}^n)$$
$$\operatorname{Op_W}(b) : \mathcal{S}(\mathbb{R}^n) \longrightarrow \mathcal{S}(\mathbb{R}^n)$$

or when

$$\operatorname{Op_W}(a) : L^2(\mathbb{R}^n) \longrightarrow L^2(\mathbb{R}^n)$$
$$\operatorname{Op_W}(b) : L^2(\mathbb{R}^n) \longrightarrow \mathcal{S}(\mathbb{R}^n).$$

Writing the compose in Weyl form

$$\widehat{C} = \left(\frac{1}{2\pi\hbar}\right)^n \int_{\mathbb{R}^{2n}} c_\sigma(z)\widehat{D}(z)dz,$$

we set out to determine the symbol c (or, equivalently, its symplectic Fourier transform c_σ).

Proposition 70. *Assume the product* $\widehat{C} = \widehat{A}\widehat{B}$ *is well-defined and write* $\widehat{C} = \operatorname{Op_W}(c)$. (i) *The symplectic Fourier transform of c is given by*

$$c_\sigma(z) = \left(\frac{1}{2\pi\hbar}\right)^{2n} \int_{\mathbb{R}^{2n}} e^{\frac{i}{2\hbar}\sigma(z,z')} a_\sigma(z - z')b_\sigma(z')dz' \tag{5.16}$$

and (ii) *the symbol c is given by*

$$c(z) = \left(\frac{1}{4\pi\hbar}\right)^{2n} \int_{\mathbb{R}^{4n}} e^{\frac{i}{2\hbar}\sigma(z',z'')} a\left(z + \frac{1}{2}z'\right) b\left(z - \frac{1}{2}z''\right) dz' dz''. \tag{5.17}$$

Proof. (i) Writing the operators $\widehat{A}$ and $\widehat{B}$ in the usual form

$$\widehat{A} = \left(\frac{1}{2\pi\hbar}\right)^{n} \int_{\mathbb{R}^{2n}} a_\sigma(z_0)\widehat{D}(z_0) dz_0$$

$$\widehat{B} = \left(\frac{1}{2\pi\hbar}\right)^{n} \int_{\mathbb{R}^{2n}} b_\sigma(z_1)\widehat{D}(z_1) dz_1,$$

we have, using the property (2.6) of displacement operators,

$$\begin{aligned}\widehat{D}(z_0)\widehat{B} &= \left(\frac{1}{2\pi\hbar}\right)^{n} \int_{\mathbb{R}^{2n}} b_\sigma(z_1)\widehat{D}(z_0)\widehat{D}(z_1) dz_1 \\ &= \left(\frac{1}{2\pi\hbar}\right)^{n} \int_{\mathbb{R}^{2n}} e^{\frac{i}{2\hbar}\sigma(z_0,z_1)} b_\sigma(z_1)\widehat{D}(z_0 + z_1) dz_1,\end{aligned}$$

and hence

$$\widehat{A}\widehat{B} = \left(\frac{1}{2\pi\hbar}\right)^{2n} \int_{\mathbb{R}^{4n}} e^{\frac{i}{2\hbar}\sigma(z_0,z_1)} a_\sigma(z_0) b_\sigma(z_1)\widehat{D}(z_0 + z_1) dz_0 dz_1.$$

Setting $z = z_0 + z_1$ and $z' = z_1$, this can be written

$$\widehat{A}\widehat{B} = \left(\frac{1}{2\pi\hbar}\right)^{2n} \int_{\mathbb{R}^{2n}} \left(\int_{\mathbb{R}^{2n}} e^{\frac{i}{2\hbar}\sigma(z,z')} a_\sigma(z - z') b_\sigma(z') dz' \right) \widehat{D}(z) dz$$

hence (5.16). (ii) Formula (5.17) is proven in a similar way, writing

$$\widehat{A} = \left(\frac{1}{\pi\hbar}\right)^{n} \int_{\mathbb{R}^{2n}} a(z_0)\widehat{R}(z_0) dz_0$$

$$\widehat{B} = \left(\frac{1}{\pi\hbar}\right)^{n} \int_{\mathbb{R}^{2n}} b(z_1)\widehat{R}(z_1) dz_1$$

and using the properties of the reflection operators $\widehat{R}(z)$. □

5.5 Comments and references

There is a multitude of books and treatises on the Weyl transform, so it is an impossible task to cite them all. Here are a few of our personal preferences: Folland [28]; his approach is particularly well-suited to an audience having a serious background in time-frequency analysis; K. Gröchenig's book [40] is also designed for a TFA public, but is very complete, and nevertheless easy to read. Historically, the first mathematically rigorous treatment was given by Hörmander in [45], who developed the ideas contained in his earlier paper [46]. The Shubin symbol classes were introduced by Shubin in [66]. They have the advantage over the usual standard Hörmander symbol classes to be global, in the sense that the coordinates x and p are placed on equal levels, while the Hörmander classes privilege the growth properties of the symbols in the p coordinates. This difference of treatment is due to the specificity of the problems addressed by these authors. While Hörmander had in mind the study of partial differential equations

$$\sum_{|\alpha|} a_\alpha(x) D_x^\alpha u(x) = f(x),$$

Shubin wanted to develop a tool more suitable to quantum operators, which usually arise from Hamiltonian functions where x and p are placed on equal footing. For a proof of Schwartz's kernel theorem, see for instance Gröchenig [40].

6 The Cohen class

6.1 Definition of the Cohen class

Definition 71. Let $Q : \mathcal{S}(\mathbb{R}^n) \times \mathcal{S}(\mathbb{R}^n) \longrightarrow \mathcal{S}(\mathbb{R}^{2n})$ be a sesquilinear form. We say that Q belongs to the Cohen class if there exists a tempered distribution $\theta \in \mathcal{S}'(\mathbb{R}^{2n})$ (the "Cohen kernel") such that

$$Q(\psi, \phi) = W(\psi, \phi) * \theta \tag{6.1}$$

for all $\psi, \phi \in \mathcal{S}(\mathbb{R}^n)$. When $\psi = \phi$, one writes $Q\psi = Q(\psi, \psi)$.

Observe that it follows from formula (3.23) that the definition (6.1) is equivalent to

$$F_\sigma Q(\psi, \phi) = (2\pi\hbar)^n A(\psi, \phi) F_\sigma \theta \tag{6.2}$$

where F_σ is the symplectic Fourier transform and $A(\psi, \phi)$ the cross-ambiguity function.

Choosing $\theta = \delta$, we have $Q(\psi, \phi) = W(\psi, \phi)$, hence the cross-Wigner transform trivially belongs to the Cohen class. Choosing

$$\theta(z) = (\pi\hbar)^{-n} e^{-\frac{1}{\hbar}|z|^2},$$

we get the so-called the Husimi distribution.

The following result gives a sufficient condition for a sesquilinear form Q to belong to Cohen's class when it takes its values in $\mathcal{S}(\mathbb{R}^{2n})$.

Proposition 72. *Let $Q : \mathcal{S}(\mathbb{R}^n) \times \mathcal{S}(\mathbb{R}^n) \longrightarrow \mathcal{S}(\mathbb{R}^{2n})$ be a sesquilinear form such that*

$$Q(\widehat{D}(z_0)\psi)(z) = Q\psi(z - z_0) \tag{6.3}$$

$$|Q(\psi, \phi)(0,0)| \leq \|\psi\| \, \|\phi\| \tag{6.4}$$

for all ψ, ϕ in $\mathcal{S}(\mathbb{R}^n)$, then Q belongs to the Cohen class.

Proof. The condition (6.4) means that the sesquilinear form Q is bounded. It follows, using Riesz's representation theorem, that there exists a bounded operator $\widehat{A}$ on $L^2(\mathbb{R}^n)$ such that

$$Q(\psi, \phi)(0,0) = (\widehat{A}\psi|\phi)_{L^2}.$$

Using the translation property (6.3), we then have

$$Q\psi(z_0) = Q(\widehat{D}(-z_0)\psi)(0) = (\widehat{A}\widehat{D}(-z_0)\psi, \widehat{D}(-z_0)\psi)_{L^2}.$$

In view of Schwartz's kernel theorem, there exists a distribution $K \in \mathcal{S}'(\mathbb{R}^n \times \mathbb{R}^n)$ such that

$$(\phi|\widehat{A}\psi)_{L^2} = \langle K, \psi \otimes \overline{\phi} \rangle$$

https://doi.org/10.1515/9783110722772-006

for all $(\psi, \phi) \in \mathcal{S}(\mathbb{R}^n) \times \mathcal{S}(\mathbb{R}^n)$. We thus have

$$Q\psi(z_0) = \langle K, \widehat{D}(-z_0)\psi \otimes \overline{\widehat{D}(-z_0)\psi}\rangle,$$

which we write formally in integral form as

$$Q\psi(z_0) = \int_{\mathbb{R}^{2n}} K(x,y)\widehat{D}(-z_0)\psi(x)\overline{\widehat{D}(-z_0)\psi(y)}dxdy.$$

By the definition of $\widehat{D}(z_0)$, we have

$$\widehat{D}(-z_0)\psi(x) = e^{\frac{i}{\hbar}(-p_0\cdot x - \frac{1}{2}p_0\cdot x_0)}\psi(x + x_0),$$

and hence

$$Q\psi(z_0) = \int_{\mathbb{R}^{2n}} e^{-\frac{i}{\hbar}p_0(x-y)}K(x,y)\psi(x+x_0)\overline{\psi(y+x_0)}dxdy. \tag{6.5}$$

On the other hand, for every $\theta \in \mathcal{S}'(\mathbb{R}^{2n})$, we have

$$(W\psi * \theta)(z_0) = \int_{\mathbb{R}^{2n}} W\psi(z_0 - z)\theta(z)dz,$$

hence, in view of the definition of the Wigner transform,

$$(W\psi * \theta)(z_0) = \left(\frac{1}{2\pi\hbar}\right)^n \int_{\mathbb{R}^{3n}} e^{-\frac{i}{\hbar}(p_0-p)\cdot y'}\psi\left(x_0 - x' + \frac{1}{2}y'\right) \times \overline{\psi\left(x_0 - x' - \frac{1}{2}y'\right)}\theta(x',p')dpdx'dy'.$$

Calculating the integral in p, we get

$$(W\psi * \theta)(z_0) = \left(\frac{1}{2\pi\hbar}\right)^{n/2} \int_{\mathbb{R}^{3n}} F_2^{-1}\theta(x',y')e^{-\frac{i}{\hbar}p_0\cdot y'}\psi\left(x_0 - x' + \frac{1}{2}y'\right) \times \overline{\psi\left(x_0 - x' - \frac{1}{2}y'\right)}\theta(x',p')dx'dy'$$

where F_2^{-1} is the inverse Fourier transform in y'. Making the change of variables $x' = -\frac{1}{2}(x+y)$ and $y' = x - y$, we have $dx'dy' = dxdy$, so the equality above becomes

$$(W\psi * \theta)(z_0) = \left(\frac{1}{2\pi\hbar}\right)^{n/2} \int_{\mathbb{R}^{2n}} F_2^{-1}\theta(x,x-y)e^{-\frac{i}{\hbar}p_0\cdot(x-y)}\psi(x+x_0)\overline{\psi(y+x_0)}dxdy.$$

Comparison with (6.5) shows that $Q\psi = W\psi * \theta$ where θ is defined by the partial inverse Fourier transform

$$K(x,y) = \left(\frac{1}{2\pi\hbar}\right)^{n/2} F_2^{-1}\theta(x, x-y),$$

and hence the functions

$$\theta(x,p) = (2\pi\hbar)^{n/2} \int_{\mathbb{R}^n} e^{-\frac{i}{\hbar}p\cdot y} K(x, x-y)dy$$

and $F\theta(0,p)$ are not constant (they are Gaussians). □

6.2 A Moyal identity

We are going to show that a Moyal-type identity is satisfied by Q provided that the Fourier transform of the Cohen kernel has a constant modulus.

Proposition 73. *Let Q be an element of the Cohen class. We have the Moyal identity*

$$(Q(\psi,\phi)|Q(\psi',\phi'))_{L^2(\mathbb{R}^{2n})} = \left(\frac{1}{2\pi\hbar}\right)^n (\psi|\psi')_{L^2}\overline{(\phi|\phi')_{L^2}} \tag{6.6}$$

if and only the Fourier transform $F\theta$ satisfies

$$|F\theta(z)| = (2\pi\hbar)^{-n} \tag{6.7}$$

for some real function φ.

Proof. In view of Moyal's identity for the cross-Wigner transform, it is equivalent to prove that

$$(Q(\psi,\phi)|Q(\psi',\phi'))_{L^2(\mathbb{R}^{2n})} = (W(\psi,\phi)|W(\psi',\phi'))_{L^2(\mathbb{R}^{2n})} \tag{6.8}$$

for all ψ, ϕ, ψ', ϕ' in $\mathcal{S}(\mathbb{R}^n)$ if and only if $|F\theta(z)| = (2\pi\hbar)^n$; applying Plancherel's formula to both sides of this equality, this condition is in fact equivalent to

$$(FQ(\psi,\phi)|FQ(\psi',\phi'))_{L^2(\mathbb{R}^{2n})} = (FW(\psi,\phi)|FW(\psi',\phi'))_{L^2(\mathbb{R}^{2n})}. \tag{6.9}$$

Since $Q(\psi,\phi) = W(\psi,\phi) * \theta$, we have

$$FQ(\psi,\phi) = (2\pi\hbar)^n FW(\psi,\phi)F\theta,$$

hence (6.9) is equivalent to

$$(2\pi\hbar)^{2n}(FW(\psi,\phi)F\theta(z)|FW(\psi',\phi')F\theta(z))_{L^2(\mathbb{R}^{2n})} = (FW(\psi,\phi)|FW(\psi',\phi'))_{L^2(\mathbb{R}^{2n})},$$

and this equality can hold if and only if $|F\theta(z)|^2 = (2\pi\hbar)^{-2n}$ for all z. □

Since condition (6.7) implies the conditions (6.11), any element of the Cohen class that satisfies Moyal's identity also satisfies the marginal conditions (6.12).

6.3 The marginal properties

Let us discuss the existence of marginal properties similar to those (3.29) satisfied by the Wigner transform.

We will say that and element Q of the Cohen class satisfies the marginal conditions if we have, for $\psi \in L^1(\mathbb{R}^n) \cap L^2(\mathbb{R}^n)$ and $F\psi \in L^1(\mathbb{R}^n) \cap L^2(\mathbb{R}^n)$,

$$\int_{\mathbb{R}^n} Q\psi(z)dp = |\psi(x)|^2 \quad \text{and} \quad \int_{\mathbb{R}^n} Q\psi(z)dx = |\widehat{\psi}(p)|^2. \tag{6.10}$$

Note that when these conditions are satisfied, we have

$$\int_{\mathbb{R}^{2n}} Q\psi(z)dz = 1$$

when $\|\psi\|_{L^2} = 1$.

Here is a necessary and sufficient condition:

Proposition 74. *Let $\psi \in L^1(\mathbb{R}^n) \cap L^2(\mathbb{R}^n)$ and $F\psi \in L^1(\mathbb{R}^n) \cap L^2(\mathbb{R}^n)$. Then $Q\psi = W\psi * \theta$ satisfies the marginal properties*

$$\int_{\mathbb{R}^n} Q\psi(z)dp = |\psi(x)|^2, \quad \int_{\mathbb{R}^n} Q\psi(z)dx = |F\psi(p)|^2$$

if and only if the Fourier transform $F\theta$ exists and is such that:

$$F\theta(x, 0) = F\theta(0, p) = (2\pi\hbar)^{-n}. \tag{6.11}$$

Proof. Let us prove the more general intermediary result: If the partial mappings $x \longmapsto \theta(x, p)$ and $p \longmapsto \theta(x, p)$ are integrable, then

$$\int_{\mathbb{R}^n} Q\psi(z)dp = (|\psi|^2 * \alpha)(x), \quad \int_{\mathbb{R}^n} Q\psi(z)dx = (|F\psi|^2 * \beta)(p) \tag{6.12}$$

where the functions α and β are defined by

$$\alpha(x) = \int_{\mathbb{R}^n} \theta(x, p)dp, \quad \beta(p) = \int_{\mathbb{R}^n} \theta(x, p)dx.$$

In view of the first marginal property (3.30) satisfied by the Wigner distribution, we have

$$\int_{\mathbb{R}^n} W\psi(x - x', p - p')dp = |\psi(x - x')|^2,$$

and hence, using Fubini's theorem,

$$\begin{aligned}\int_{\mathbb{R}^n} Q\psi(z)dp &= \int_{\mathbb{R}^n}\Big(\int_{\mathbb{R}^{2n}} W\psi(z-z')\theta(z')dz'\Big)dp \\ &= \int_{\mathbb{R}^{2n}}\Big(\int_{\mathbb{R}^n} W\psi(z-z')dp\Big)\theta(z')dz' \\ &= \int_{\mathbb{R}^n} |\psi(x-x')|^2\Big(\int_{\mathbb{R}^n}\theta(z')dp'\Big)dx'\end{aligned}$$

which yields the first formula (6.12). The second formula is proven in a similar way using the second marginal property (3.31) of the Wigner transform. (ii) It suffices to show that the conditions (6.11) imply that $\alpha(x) = \delta(x)$ and $\beta(p) = \delta(p)$. Let $\widehat{\theta}$ be the usual Fourier transform of the kernel θ; we have

$$\begin{aligned}F\theta(x,0) &= \Big(\frac{1}{2\pi\hbar}\Big)^n \int_{\mathbb{R}^{2n}} e^{-\frac{i}{\hbar}xx'}\theta(x',p')dp'dx' \\ &= \Big(\frac{1}{2\pi\hbar}\Big)^n \int_{\mathbb{R}^n} e^{-\frac{i}{\hbar}xx'}\Big(\int_{\mathbb{R}^n}\theta(x',p')dp'\Big)dx' \\ &= \Big(\frac{1}{2\pi\hbar}\Big)^n \int_{\mathbb{R}^n} e^{-\frac{i}{\hbar}xx'}\alpha(x')dx',\end{aligned}$$

hence the condition $F\theta(x,0) = (2\pi\hbar)^{-n}$ is equivalent to $\alpha(x) = \delta(x)$. Similarly, we have $F\theta(0,p) = (2\pi\hbar)^{-n}$. □

Remark 75. In particular, if Q satisfies a Moyal identity, then it also satisfies the marginal conditions.

6.4 The τ-Wigner transform

Here we study a very particular member of the Cohen class. It will be instrumental in our study of Born–Jordan quantization later in this book.

6.4.1 Definition

In what follows, τ is a real parameter.

Definition 76. The element W_τ of the Cohen class corresponding to the Cohen kernel

$$\theta_\tau(z) = \begin{cases} \big(\frac{1}{2\pi\hbar}\big)^n \frac{2^n}{|2\tau-1|^n} e^{\frac{2i}{\hbar(2\tau-1)}p\cdot x} & \text{for } \tau \neq \frac{1}{2} \\ \delta(z) & \text{for } \tau = \frac{1}{2} \end{cases} \tag{6.13}$$

is called the τ-Wigner transform. We call θ_τ the Shubin τ-kernel.

Of course $W_{1/2}(\psi,\phi)$ is the usual Wigner function $W(\psi,\phi)$.
Let give an explicit description of $W_\tau(\psi,\phi)$ for $\tau \neq \frac{1}{2}$.

Proposition 77. *Let $(\psi,\phi) \in L^2(\mathbb{R}^n) \times L^2(\mathbb{R}^n)$. We have, for $\tau \neq \frac{1}{2}$,*

$$W_\tau(\psi,\phi)(z) = \left(\frac{1}{2\pi\hbar}\right)^n \int_{\mathbb{R}^n} e^{-\frac{i}{\hbar}py}\psi(x+\tau y)\overline{\phi(x-(1-\tau)y)}dy. \tag{6.14}$$

Proof. It suffices to give the proof for $(\psi,\phi) \in \mathcal{S}(\mathbb{R}^n) \times \mathcal{S}(\mathbb{R}^n)$; the expression (6.14) extends to the general case as for the usual cross-Wigner transform. By definition of the convolution product, $W(\psi,\phi) * \theta_\tau$ is given by the absolutely convergent integral

$$W(\psi,\phi) * \theta_\tau(z) = \int_{\mathbb{R}^{2n}} W(\psi,\phi)(z-z')\theta_\tau(z')dz'.$$

Expanding the integrand using formula (3.4) for the cross-Wigner transform, we get

$$W(\psi,\phi)(z-z')\theta_\tau(z') = \left(\frac{1}{2\pi\hbar}\right)^{2n} \frac{2^n}{|2\tau-1|^n} \int_{\mathbb{R}^n} e^{-\frac{i}{\hbar}py} e^{\frac{i}{\hbar}p'(y+\frac{2x'}{2\tau-1})}$$
$$\times \psi\left(x-x'+\frac{1}{2}y\right)\overline{\phi\left(x-x'-\frac{1}{2}y\right)}dy.$$

Using the identity

$$\int_{\mathbb{R}^n} e^{\frac{i}{\hbar}p'(y+\frac{2x'}{2\tau-1})}dp' = (2\pi\hbar)^n\delta\left(y+\frac{2x'}{2\tau-1}\right)$$

and integrating with respect to p' yields

$$\int_{\mathbb{R}^n} W(\psi,\phi)(z-z')\theta_\tau(z')dp' = \frac{2^n}{|2\tau-1|^n}\left(\frac{1}{2\pi\hbar}\right)^n \int_{\mathbb{R}^n} e^{-\frac{i}{\hbar}py}\delta\left(y+\frac{2x'}{2\tau-1}\right)$$
$$\times \psi\left(x-x'+\frac{1}{2}y\right)\overline{\phi\left(x-x'-\frac{1}{2}y\right)}dy$$

that is

$$\int_{\mathbb{R}^n} W(\psi,\phi)(z-z')\theta_\tau(z')dp'$$
$$= \frac{2^n}{|2\tau-1|^n}\left(\frac{1}{2\pi\hbar}\right)^n \int_{\mathbb{R}^n} e^{-\frac{i}{\hbar}py}\delta\left(y+\frac{2x'}{2\tau-1}\right)\psi(x+\tau y)\overline{\phi(x-(1-\tau)y)}dy$$

where we have used the equality

$$\delta\left(y+\frac{2x'}{2\tau-1}\right)\psi\left(x-x'+\frac{1}{2}y\right)\overline{\phi\left(x-x'-\frac{1}{2}y\right)}$$
$$=\delta\left(y+\frac{2x'}{2\tau-1}\right)\psi(x+\tau y)\overline{\phi(x-(1-\tau)y)}dy.$$

Integrating now in x' and taking the identity

$$\int_{\mathbb{R}^n}\delta\left(y+\frac{2x'}{2\tau-1}\right)dx'=\left(\frac{|2\tau-1|}{2}\right)^n$$

into account, we finally get

$$W(\psi,\phi)*\theta_\tau(z)=\left(\frac{1}{2\pi\hbar}\right)^n\int_{\mathbb{R}^n}e^{-\frac{i}{\hbar}py}\psi(x+\tau y)\overline{\phi(x-(1-\tau)y)}dy$$

which is (6.14). □

Notice that, putting $\tau=\frac{1}{2}$ on the right-hand-side of (6.14), one recovers the standard cross-Wigner transform $W(\psi,\phi)$. This can be seen as a limiting case of the definition since $\theta_\tau\to\delta$ when $\tau\to\frac{1}{2}$.

If $\tau=0$, we get

$$W_0(\psi,\phi)(z)=\left(\frac{1}{2\pi\hbar}\right)^{n/2}e^{-\frac{i}{\hbar}px}\psi(x)\overline{F\phi}(p)\ :$$

W_0 is called the Rihaczek–Kirkwood distribution in time-frequency analysis and signal theory. For $\tau=1$, one gets the so-called dual Rihaczek–Kirkwood distribution.

It is easily verified using (6.14) that we have the complex conjugation formula

$$W_\tau(\phi,\psi)=\overline{W_{1-\tau}(\psi,\phi)}.$$

6.4.2 Properties

Proposition 78. (i) *The Fourier transform of the Shubin τ-kernel is given for $\tau\neq\frac{1}{2}$ by*

$$F\theta_\tau(z)=\left(\frac{1}{2\pi\hbar}\right)^n e^{-\frac{i(2\tau-1)}{2\hbar}p\cdot x}. \tag{6.15}$$

(ii) *The cross τ-Wigner transform satisfies the Moyal identity*

$$(W_\tau(\psi,\phi)|W_\tau(\psi',\phi'))_{L^2(\mathbb{R}^{2n})}=\left(\frac{1}{2\pi\hbar}\right)^n(\psi|\psi')_{L^2}\overline{(\phi|\phi')_{L^2}} \tag{6.16}$$

for all $(\psi, \phi) \in L^2(\mathbb{R}^n) \times L^2(\mathbb{R}^n)$; (iii) *The* τ*-Wigner transform satisfies the marginal conditions*

$$\int_{\mathbb{R}^n} W_\tau\psi(z)dp = |\psi(x)|^2, \quad \int_{\mathbb{R}^n} W_\tau\psi(z)dx = |F\psi(p)|^2 \tag{6.17}$$

for all $\psi \in L^1(\mathbb{R}^n) \cap L^2(\mathbb{R}^n)$ *such that* $F\psi \in L^1(\mathbb{R}^n) \cap L^2(\mathbb{R}^n)$.

Proof. Let us prove formula (6.15); it implies that $|F\theta_\tau(z)| = (2\pi\hbar)^{-n}$ and properties (ii) and (iii) will then follow in view of Proposition 73 and Remark 75 following the proof of Proposition 74. The Fourier transform of θ_τ is given by

$$F\theta_\tau(z) = \left(\frac{1}{2\pi\hbar}\right)^{2n} \frac{2^n}{|2\tau - 1|^n} \int_{\mathbb{R}^{2n}} e^{-\frac{i}{\hbar}z\cdot z'} e^{\frac{2i}{\hbar(2\tau-1)}p'\cdot x'} dp'dx'.$$

We have, since $z \cdot z' = x \cdot x' + p \cdot p'$,

$$\begin{aligned}\int_{\mathbb{R}^{2n}} e^{-\frac{i}{\hbar}z\cdot z'} e^{\frac{2i}{\hbar(2\tau-1)}p'\cdot x'} dz' &= \int_{\mathbb{R}^n} e^{-\frac{i}{\hbar}x\cdot x'} \left[\int_{\mathbb{R}^n} e^{-\frac{i}{\hbar}p'\cdot(p-\frac{2}{2\tau-1}x')} dp'\right] dx' \\ &= (2\pi\hbar)^n \int_{\mathbb{R}^n} e^{-\frac{i}{\hbar}x\cdot x'} \delta\left(p - \frac{2}{2\tau-1}x'\right) dx' \\ &= \frac{|2\tau-1|^n}{2^n}(2\pi\hbar)^n e^{-\frac{i(2\tau-1)}{2\hbar}p\cdot x},\end{aligned}$$

hence formula (6.15). □

It is remarkable that Definition 71 of the Cohen class can be reformulated using the τ-Wigner transform for any fixed value of τ:

Proposition 79. *Letting* $\tau \in [0, 1]$, *then every* $Q(\psi, \phi) = W(\psi, \phi) * \theta$ *in the Cohen class can be written*

$$Q(\psi, \phi) = W_\tau(\psi, \phi) * f_\tau \tag{6.18}$$

for a suitable $f_\tau \in \mathcal{S}'(\mathbb{R}^n)$.

Proof. In view of formula (6.15), we have $F\theta_\tau F\theta_{1-\tau} = (2\pi\hbar)^{-2n}$, hence

$$F(\theta_\tau * \theta_{1-\tau}) = (2\pi\hbar)^n F\theta_\tau F\theta_{1-\tau} = (2\pi\hbar)^{-n}$$

so that, using the Fourier inversion formula $\theta_\tau * \theta_{1-\tau} = \delta$, it follows that $Q(\psi, \phi) = W(\psi, \phi) * \theta$ can be written

$$\begin{aligned}Q(\psi, \phi) &= (W(\psi, \phi) * \theta) * (\theta_\tau * \theta_{1-\tau}) \\ &= (W(\psi, \phi) * \theta_\tau) * (\theta * \theta_{1-\tau})\end{aligned}$$

whence (6.18) with $f_\tau = \theta * \theta_{1-\tau}$. For this construction to make sense, we must however prove that f_τ exists and that $f_\tau \in \mathcal{S}'(\mathbb{R}^n)$, which is not *a priori* clear. Since $\theta * \theta_{1-\tau} = (2\pi\hbar)^n F^{-1}(F\theta F\theta_{1-\tau})$, it is sufficient to show that $F\theta F\theta_{1-\tau} \in \mathcal{S}'(\mathbb{R}^n)$. Now $F\theta \in \mathcal{S}'(\mathbb{R}^n)$ and in view of (6.15)

$$F\theta_{1-\tau} = \left(\frac{1}{2\pi\hbar}\right)^n e^{\frac{i(2\tau-1)}{2\hbar}p\cdot x},$$

which is a C^∞ function with polynomially bounded derivatives; therefore, $F\theta F\theta_{1-\tau} \in \mathcal{S}'(\mathbb{R}^n)$. □

6.5 The Born–Jordan kernel

The choice $\theta = \delta$ for the Cohen kernel is the standard one when one deals with the usual Wigner formalism. It leads, as we will see, to Weyl quantization of symbols. Another very important choice that seems to be a priori pulled out of thin air is the following:

Definition 80. The Cohen kernel θ_{BJ} obtained by averaging the Shubin τ-kernels over $\tau \in [0,1]$, that is

$$\theta_{\mathrm{BJ}} = \int_0^1 \theta_\tau d\tau, \tag{6.19}$$

is called the "Born–Jordan kernel".

We shall not attempt to calculate the integral (6.19) here, but rather describe θ_{BJ} by giving its (symplectic) Fourier transform.

Proposition 81. *The symplectic Fourier transform of the Born–Jordan kernel is given by*

$$F_\sigma\theta_{\mathrm{BJ}}(z) = \left(\frac{1}{2\pi\hbar}\right)^n \frac{\sin(px/2\hbar)}{px/2\hbar} \tag{6.20}$$

with the convention $\sin(0)/0 = 1$.

Proof. In view of formula (6.15), the Fourier transform of the Shubin τ-kernel θ_τ is given by

$$F\theta_\tau(z) = \left(\frac{1}{2\pi\hbar}\right)^n e^{-\frac{i(2\tau-1)}{2\hbar}p\cdot x}. \tag{6.21}$$

Integrating both sides of this equality in $\tau \in [0,1]$ yields

$$\int_0^1 F\theta_\tau(z)d\tau = \left(\frac{1}{2\pi\hbar}\right)^n \int_0^1 e^{-\frac{i(2\tau-1)}{2\hbar}p\cdot x}d\tau$$

$$= \left(\frac{1}{2\pi\hbar}\right)^n \frac{\sin(px/2\hbar)}{px/2\hbar}.$$

Formula (6.20) follows since

$$F_\sigma \theta_{\mathrm{BJ}}(x,p) = F\theta_{\mathrm{BJ}}(p,-x). \qquad \square$$

Formula (6.20) can be rewritten for short

$$F_\sigma \theta_{\mathrm{BJ}}(z) = \left(\frac{1}{2\pi\hbar}\right)^n \operatorname{sinc}(px/2\hbar)$$

where $\operatorname{sinc}\alpha = \sin\alpha/\alpha$ for $\alpha \neq 0$ is the *sinus cardinalis* function well-known from sampling theory.

The element of the Cohen class associated with the Born–Jordan kernel

$$W_{\mathrm{BJ}}(\psi,\phi) = W(\psi,\phi) * \theta_{\mathrm{BJ}}$$

satisfies the marginal properties in Proposition 74 because

$$F\theta_{\mathrm{BJ}}(x,0) = F\theta_{\mathrm{BJ}}(0,p) = (2\pi\hbar)^{-n}.$$

Therefore, if $\psi \in L^1(\mathbb{R}^n) \cap L^2(\mathbb{R}^n)$ and $F\psi \in L^1(\mathbb{R}^n) \cap L^2(\mathbb{R}^n)$, we will have

$$\int_{\mathbb{R}^n} W_{\mathrm{BJ}}\psi(z)dp = |\psi(x)|^2 \tag{6.22}$$

$$\int_{\mathbb{R}^n} W_{\mathrm{BJ}}\psi(z)dx = |\widehat{\psi}(p)|^2. \tag{6.23}$$

Notice that the Moyal identity does not hold for $W_{\mathrm{BJ}}(\psi,\phi)$ since we have $|F\theta_{\mathrm{BJ}}(z)| \neq (2\pi\hbar)^{-n}$ (Proposition 73).

6.6 Comments and references

The notion of quasi-distribution Q goes back to the seminal work of Leon Cohen [14, 15]. The terminology "Cohen class" was apparently introduced by Gröchenig in [40]. In the proof of Proposition 72, we have been following ([31, 40]) with some minor modifications. Boggiatto and his collaborators have studied the τ-dependent transforms $W_\tau(\psi,\phi)$. Proposition 79 is due to Boggiatto, De Donno, and Oliaro [7].

7 Born–Jordan quantization

7.1 Physical origins

To understand what Born–Jordan operators are about one has to go back to the early years of quantum mechanics (around 1925), when the notion of quantization was first discussed in depth by the physicists Max Born and Pascual Jordan following ideas of Werner Heisenberg. These investigations quickly led to the so-called "ordering problem" for operators. In its most basic form the question was the following: if we accept that the quantum version of the variables x and p should be the non-commuting operators $\widehat{x}$ and $\widehat{p}$ where $\widehat{x}$ is multiplication by x and $\widehat{p} = -i\hbar\partial_x$, what should then the quantization rule for the product xp be? Should it be $\widehat{x}\widehat{p}$ or $\widehat{p}\widehat{x}$? Or should it be something else? Consensus that the "right" choice should be the average $\frac{1}{2}(\widehat{x}\widehat{p} + \widehat{p}\widehat{x})$ was quickly reached, using symmetry considerations (Erwin Schrödinger had already proposed this choice in previous work related to his eponymous equation). Born and Jordan thereafter proposed a more general rule for "quantizing" any monomial in the variables x and p: they suggested to associate to the function $a(x,p) = x^m p^\ell$ the differential operator

$$\widehat{A}_{\mathrm{BJ}} = \frac{1}{\ell+1}\sum_{k=0}^{m} \widehat{p}^k \widehat{x}^\ell \widehat{p}^{\ell-k}. \tag{7.1}$$

Their definition seemed to be pulled out of thin air. It is actually a *rigorous* consequence of Born and Jordan's attempt to "quantize" Hamiltonian mechanics, and replace the Hamilton equations of motion

$$\frac{d\widehat{x}}{dt} = \frac{\partial H}{\partial \widehat{p}}(\widehat{x},\widehat{p}), \qquad \frac{d\widehat{p}}{dt} = -\frac{\partial H}{\partial \widehat{x}}(\widehat{x},\widehat{p})$$

with a "quantum" counterpart, postulated to be

$$\frac{dx}{dt} = \frac{\partial H}{\partial p}(x,p), \qquad \frac{dp}{dt} = -\frac{\partial H}{\partial x}(x,p).$$

For this reason it might very well the physically correct quantization scheme, as opposed to the Weyl quantization we studied in Chapter 5; the latter leads to the quantization rule

$$\widehat{A}_{\mathrm{Weyl}} = \frac{1}{2^\ell}\sum_{k=0}^{\ell} \binom{\ell}{k} \widehat{p}^{\ell-k} \widehat{x}^m \widehat{p}^k \tag{7.2}$$

for the monomial $a(x,p) = x^m p^\ell$; we have $\widehat{A}_{\mathrm{BJ}} \neq \widehat{A}_{\mathrm{Weyl}}$ as soon as $m + \ell > 2$. It turns out that Weyl quantization quickly superseded Born and Jordan's rule for historical reasons. The attractiveness of Weyl quantization has several origins; it enjoys the symplectic covariance property which in a sense reproduces the canonical invariance of

https://doi.org/10.1515/9783110722772-007

Hamilton's equation. We are going to see that the ideas of Born and Jordan can be very easily and elegantly interpreted in terms of a certain element of the Cohen class, defined by what we called the Born–Jordan kernel in last chapter.

7.2 Algebraic motivation

7.2.1 Weyl, Shubin, and Born–Jordan

Let a be, as above, the monomial $a(x,p) = x^m p^\ell$ where m and ℓ are non-negative integers and $\widehat{x}$ and $\widehat{p}$ be differential operators $\mathcal{S}(\mathbb{R}) \longrightarrow \mathcal{S}(\mathbb{R})$; we choose as above $\widehat{x} =$ multiplication by x and $\widehat{p} = -i\hbar\partial_x$. Let τ be a real parameter and define the *Shubin ordering* by associating to a the differential operator

$$\widehat{A}_\tau = \sum_{k=0}^{\ell} \binom{\ell}{k} (1-\tau)^k \tau^{\ell-k} \widehat{p}^k \widehat{x}^m \widehat{p}^{\ell-k}.$$

Setting $\tau = 0$ we get the operator $\widehat{A}_0 = \widehat{p}^\ell \widehat{x}^m$ and, setting $\tau = 1$, $\widehat{A}_0 = \widehat{x}^m \widehat{p}^\ell$ (these orderings are called in the literature "anti-normal ordering" and "normal ordering", respectively). More interesting is the choice $\tau = \frac{1}{2}$ which leads to the Weyl ordering (7.2):

$$\widehat{A}_{1/2} = \widehat{A}_{\mathrm{Weyl}} = \frac{1}{2^\ell} \sum_{k=0}^{\ell} \binom{\ell}{k} \widehat{p}^{\ell-k} \widehat{x}^m \widehat{p}^k.$$

While no choice of τ leads to the Born–Jordan ordering (7.1), the latter is however obtained by averaging the mapping $\tau \longmapsto \widehat{A}_\tau$ over the interval $[0,1]$. In fact, using the identity

$$\int_0^1 (1-\tau)^k \tau^{\ell-k} d\tau = \frac{k!(\ell-k)!}{(\ell+1)!}$$

familiar from the theory of the beta function we get

$$\binom{\ell}{k} \int_0^1 (1-\tau)^k \tau^{\ell-k} d\tau = \frac{1}{\ell+1}$$

so that we have

$$\widehat{A}_{\mathrm{BJ}} = \int_0^1 \widehat{A}_\tau d\tau$$

as claimed.

In the rest of this chapter we will use this "averaging" procedure to define and study Born–Jordan operators in full generality.

7.2.2 A remark on the quantization of monomials

Our search for quantizations of the monomials $a(x,p) = x^m p^\ell$ can be placed in a more axiomatic setting by introducing a set of axioms. Let $\mathbb{C}[x,p]$ be the complex vector space of all monomial functions. We define a quantization of $\mathbb{C}[x,p]$ as being a linear mapping

$$\mathrm{Op} : \mathbb{C}[x,p] \longrightarrow \mathbb{C}[\widehat{x},\widehat{p}]$$

such that:

(A1) $\mathrm{Op}(1) = 1$ (the identity), $\mathrm{Op}(x) = \widehat{x}$ and $\mathrm{Op}(p) = \widehat{p}$;

(A2) $\mathrm{Op}(a)$ is self-adjoint if $a \in \mathbb{C}[x,p]$ is a real polynomial;

(A3) The restricted Dirac correspondence

$$[\widehat{x}, \mathrm{Op}(a)] = i\hbar\, \mathrm{Op}(\{x,a\}) \tag{7.3}$$

$$[\widehat{p}, \mathrm{Op}(a)] = i\hbar\, \mathrm{Op}(\{p,a\}) \tag{7.4}$$

holds for every $a \in \mathbb{C}[x,p]$; in particular $[\widehat{x},\widehat{p}] = i\hbar$.

Proposition 82. *Let* $\mathrm{Op} : \mathbb{C}[x,p] \longrightarrow \mathbb{C}[\widehat{x},\widehat{p}]$ *be a quantization of monomials n the sense above. Then there exists a real function* $\chi \in C^\infty(\mathbb{R}^2)$ *with* $\chi(0) = 1$ *such that*

$$\mathrm{Op}(x^r p^s) = \sum_{\ell=0}^{\min(r,s)} \hbar^\ell \ell! f^{(\ell)}(0) \binom{s}{\ell}\binom{r}{\ell} \widehat{p}^{s-\ell}\widehat{x}^{r-\ell} \tag{7.5}$$

where $f^{(\ell)}(0)$ *is the* ℓ*-th derivative of the function* $f(x) = e^{ix/2}\chi(x)$ *at* $x = 0$.

Suppose for instance that $\chi(x) = 1$ for all x, then $f(x) = e^{ix/2}$ and $f^{(j)}(0) = (i/2)^j$ and it is easy to show using repeatedly the commutation relations $[\widehat{x},\widehat{p}] = i\hbar$ that in this case Op is just Weyl quantization: $\mathrm{Op} = \mathrm{Op}_\mathrm{W}$. Similarly, the choice $\chi(x) = \mathrm{sinc}(x/2)$ leads to $f^{(\ell)}(0) = i^\ell/(\ell+1)$ and Op is the Born–Jordan quantization: $\mathrm{Op} = \mathrm{Op}_\mathrm{W}$. We notice that the Shubin ordering does not correspond to any monomial quantization in the sense above. For details and references see the last Section of this Chapter.

7.3 Born–Jordan operators: definition(s)

It turns out that we have already developed much of the material we need to extend the algebraic considerations above to the case of arbitrary symbols.

7.3.1 Definition using the Cohen class

Let a be a function or distribution having vocation to be a symbol. Recall (formula (5.5)) that the Weyl operator $\mathrm{Op}_\mathrm{W}(a)$ can be defined in the weak sense, for $a \in \mathcal{S}'(\mathbb{R}^{2n})$

by the condition

$$\langle \mathrm{Op_W}(a)\psi, \overline{\phi}\rangle = \langle a, W(\psi,\phi)\rangle$$

where $(\psi,\phi) \in \mathcal{S}(\mathbb{R}^n) \times \mathcal{S}(\mathbb{R}^n)$.

Definition 83. Let $a \in \mathcal{S}'(\mathbb{R}^{2n})$. The Born–Jordan operator $\mathrm{Op_{BJ}}(a)$ is the only operator $\mathcal{S}(\mathbb{R}^n) \longrightarrow \mathcal{S}'(\mathbb{R}^n)$ such that

$$\langle \mathrm{Op_{BJ}}(a)\psi, \overline{\phi}\rangle = \langle a, W_{\mathrm{BJ}}(\psi,\phi)\rangle \tag{7.6}$$

for all $(\psi,\phi) \in \mathcal{S}(\mathbb{R}^n) \times \mathcal{S}(\mathbb{R}^n)$. Here $W_{\mathrm{BJ}}(\psi,\phi) = W(\psi,\phi) * \theta_{\mathrm{BJ}}$ where θ_{BJ} is the Born–Jordan kernel (Definition 80).

We will write $\widehat{A}_{\mathrm{BJ}} = \mathrm{Op_{BJ}}(a)$.

Recall that symplectic Fourier transform of the Born–Jordan kernel is defined by

$$F_\sigma\theta_{\mathrm{BJ}}(z) = \left(\frac{1}{2\pi\hbar}\right)^n \frac{\sin(px/2\hbar)}{px/2\hbar}. \tag{7.7}$$

While formula (7.6) indeed defines $\mathrm{Op_{BJ}}(a)$ in the weak sense, we must check the uniqueness statement. Assume that a second operator $\widehat{A}'_{\mathrm{BJ}} = \mathrm{Op}'_{\mathrm{BJ}}(a)$ satisfies (7.6); then we have

$$\langle (\widehat{A}_{\mathrm{BJ}} - \widehat{A}'_{\mathrm{BJ}})\psi, \overline{\phi}\rangle = 0$$

for all (ψ,ϕ) and hence $\widehat{A}_{\mathrm{BJ}} - \widehat{A}'_{\mathrm{BJ}} = 0$.

Proposition 84. *Let* $a \in \mathcal{S}(\mathbb{R}^{2n})$. (i) *We have*

$$\mathrm{Op_{BJ}}(a) = \mathrm{Op_W}(a * \theta_{\mathrm{BJ}}) \tag{7.8}$$

where θ_{BJ} *is the Born–Jordan kernel.* (ii) *The operator* $\mathrm{Op_{BJ}}(a)$ *has the harmonic decomposition*

$$\widehat{A}_{\mathrm{BJ}}\psi(x) = \left(\frac{1}{2\pi\hbar}\right)^n \int_{\mathbb{R}^{2n}} a_\sigma(z_0)\widehat{D}_{\mathrm{BJ}}(z_0)\psi(x)dz_0 \tag{7.9}$$

where $\widehat{D}_{\mathrm{BJ}}(z_0)$ *is the operator defined by*

$$\widehat{D}_{\mathrm{BJ}}(z_0) = \widehat{D}(z_0)\,\mathrm{sinc}\left(\frac{p_0x_0}{2\hbar}\right). \tag{7.10}$$

Proof. (i) In integral notation definition (7.6) reads

$$\int_{\mathbb{R}^n} \widehat{A}_{\mathrm{BJ}}\psi(x)\overline{\phi(x)}dx = \int_{\mathbb{R}^{2n}} a(z)W_{\mathrm{BJ}}(\psi,\phi)(z)dz\,;$$

using the Plancherel formula for the symplectic Fourier transform F_σ this can be rewritten

$$\int_{\mathbb{R}^n} \widehat{A}_{\mathrm{BJ}}\psi(x)\overline{\phi(x)}dx = \int_{\mathbb{R}^{2n}} F_\sigma a(z) F_\sigma W_{\mathrm{BJ}}(\psi,\phi)(-z)dz.$$

In view of the formula $F_\sigma(f * g) = (2\pi\hbar)^n F_\sigma f F_\sigma g$ we have

$$F_\sigma W_{\mathrm{BJ}}(\psi,\phi) = (2\pi\hbar)^n F_\sigma W(\psi,\phi) F_\sigma \theta_{\mathrm{BJ}}$$

and hence

$$\int_{\mathbb{R}^n} \widehat{A}_{\mathrm{BJ}}\psi(x)\overline{\phi(x)}dx = (2\pi\hbar)^n \int_{\mathbb{R}^{2n}} F_\sigma a(z) F_\sigma W(\psi,\phi)(-z) F_\sigma \theta_{\mathrm{BJ}}(-z)dz$$

that is, since $F_\sigma\theta_{\mathrm{BJ}}(-z) = F_\sigma\theta_{\mathrm{BJ}}(z)$,

$$\begin{aligned}\int_{\mathbb{R}^n} \widehat{A}_{\mathrm{BJ}}\psi(x)\overline{\phi}(x)dx &= (2\pi\hbar)^n \int_{\mathbb{R}^{2n}} F_\sigma a(z) F_\sigma \theta_{\mathrm{BJ}}(z) F_\sigma W(\psi,\phi)(-z)dz \\ &= \int_{\mathbb{R}^{2n}} F_\sigma(a * \theta_{\mathrm{BJ}})(z) F_\sigma W(\psi,\phi)(-z)dz.\end{aligned}$$

Using again Plancherel's formula we get

$$\int_{\mathbb{R}^n} \widehat{A}_{\mathrm{BJ}}\psi(x)\overline{\phi}(x)dx = \int_{\mathbb{R}^{2n}} (a * \theta_{\mathrm{BJ}})(z) W(\psi,\phi)(z)dz$$

hence formula (7.8). (ii) Taking definition (7.7) into account we have

$$F_\sigma W_{\mathrm{BJ}}(\psi,\phi)(z) = \frac{\sin(px/2\hbar)}{px/2\hbar} F_\sigma W(\psi,\phi)(z).$$

(Notice that $F_\sigma W(\psi,\phi)$ is just the cross-ambiguity function $A(\psi,\phi)$). It follows from the calculations above that we have the identity (7.9) where $\widehat{D}(z_0)$ is the usual Heisenberg displacement operator. □

Notice that the operator $\widehat{D}_{\mathrm{BJ}}(z_0)$ defined by (7.10) is not unitary on $L^2(\mathbb{R}^n)$.

7.3.2 Pseudodifferential expression

Here is an alternative definition of $\mathrm{Op}_{\mathrm{BJ}}(a)$ which links the results above to the averaging procedure introduced in the beginning of this chapter.

Recall that we defined in last chapter (formula (6.13)) the τ-Wigner transform (τ a real parameter) as the element of the Cohen class associated to the Shubin τ-kernel given by

$$\theta_\tau(z) = \left(\frac{1}{2\pi\hbar}\right)^n \frac{2^n}{|2\tau-1|^n} e^{\frac{2i}{\hbar(2\tau-1)}p\cdot x}$$

for $\tau = \frac{1}{2}$, and $\theta_{1/2}(z) = \delta(z)$. We thereafter defined (formula (6.19)) the Born–Jordan kernel θ_{BJ} by averaging the Shubin τ-kernels over the real interval $[0,1]$

$$\theta_{\mathrm{BJ}} = \int_0^1 \theta_\tau d\tau. \tag{7.11}$$

Definition 85. Let $\tau \in \mathbb{R}$ and $a \in \mathcal{S}(\mathbb{R}^{2n})$. The Shubin τ-pseudodifferential operator $A_\tau = \mathrm{Op}_\tau(a)$ is the only operator $\mathcal{S}(\mathbb{R}^n) \longrightarrow \mathcal{S}'(\mathbb{R}^n)$ such that

$$\langle \mathrm{Op}_\tau(a)\psi, \overline{\phi}\rangle = \langle a, W_\tau(\psi,\phi)\rangle \tag{7.12}$$

for all $(\psi,\phi) \in \mathcal{S}(\mathbb{R}^n) \times \mathcal{S}(\mathbb{R}^n)$.

Here $W_\tau(\psi,\phi) = W(\psi,\phi) * \theta_\tau$ is the τ-Wigner transform studied in last chapter (Definition 76).

Proposition 86. *Let $a \in \mathcal{S}(\mathbb{R}^{2n})$ and $\psi \in \mathcal{S}(\mathbb{R}^n)$.* (i) *The Shubin operator $\widehat{A}_\tau = \mathrm{Op}_\tau(a)$ is given by*

$$\widehat{A}_\tau\psi(x) = \left(\frac{1}{2\pi\hbar}\right)^n \int_{\mathbb{R}^{2n}} e^{\frac{i}{\hbar}p(x-y)} a((1-\tau)x+\tau y, p)\psi(y)dydp \tag{7.13}$$

and (ii) *The Born–Jordan operator $\widehat{A}_{BJ} = \mathrm{Op}_{\mathrm{BJ}}(a)$ is given by*

$$\widehat{A}_{BJ} = \int_0^1 \widehat{A}_\tau d\tau. \tag{7.14}$$

Proof. (i) In view of formula (6.14) of last chapter we have

$$W_\tau(\psi,\phi)(z) = \left(\frac{1}{2\pi\hbar}\right)^n \int_{\mathbb{R}^n} e^{-\frac{i}{\hbar}py}\psi(x+\tau y)\overline{\phi(x-(1-\tau)y)}dy \tag{7.15}$$

and hence

$$\langle a, W_\tau(\psi,\phi)\rangle = \left(\frac{1}{2\pi\hbar}\right)^n \int_{\mathbb{R}^{3n}} e^{-\frac{i}{\hbar}py} a(x,p)\psi(x+\tau y) \times \overline{\phi(x-(1-\tau)y)}dydpdx.$$

Performing the change of variables $x' = x - (1-\tau)y$ and $y' = x + \tau y$ this is

$$\langle a, W_\tau(\psi,\phi)\rangle = \left(\frac{1}{2\pi\hbar}\right)^n \int_{\mathbb{R}^{3n}} e^{-\frac{i}{\hbar}p(x'-y')} \times a((1-\tau)x' + \tau y', p)\psi(y')\overline{\phi(x')}d^n y' d^n p d^n x'$$

hence the equality (7.13). (ii) Formula (7.14) follows. □

Notice that a Born–Jordan operator can be written in the form of a Fourier integral operator: setting

$$b(x,y,p) = \int_0^1 a((1-\tau)x + \tau y, p)d\tau$$

we have

$$\widehat{A}_{\mathrm{BJ}}\psi(x) = \left(\frac{1}{2\pi\hbar}\right)^n \int_{\mathbb{R}^{2n}} e^{\frac{i}{\hbar}p(x-y)} b(x,y,p)\psi(y)dydp.$$

The following result in a sense “trivializes” Born–Jordan quantization; it shows that if we choose the symbol in one of the Shubin classes, then $\mathrm{Op}_{\mathrm{BJ}}(a)$ is essentially a Shubin pseudodifferential operator:

Proposition 87. (i) *If* $\widehat{A}_{\mathrm{BJ}} = \mathrm{Op}_{\mathrm{BJ}}(a)$ *with* $a \in \Gamma_\rho^m(\mathbb{R}^{2n})$ *there exists, for every* $\tau \in \mathbb{R}$, *a symbol* a_τ *belonging to the same symbol class* $\Gamma_\rho^m(\mathbb{R}^{2n})$ *such that* $\widehat{A}_{\mathrm{BJ}} = \mathrm{Op}_\tau(a_\tau)$. (ii) *Conversely, for any given symbol* $a_\tau \in \Gamma_\rho^m(\mathbb{R}^{2n})$ *there exists a symbol* $a \in \Gamma_\rho^m(\mathbb{R}^{2n})$ *such that* $\mathrm{Op}_{\mathrm{BJ}}(a) = \mathrm{Op}_\tau(a_\tau) + R$ *where* R *is an operator with integral kernel in* $\mathcal{S}(\mathbb{R}^{2n})$.

In particular, taking $\tau = \frac{1}{2}$, the operator $\widehat{A}_{\mathrm{BJ}} = \mathrm{Op}_{\mathrm{BJ}}(a)$ is a Weyl pseudodifferential operator with symbol in the same Shubin class as a.

7.4 Non-invertibility of Born–Jordan quantization

The Weyl transform, viewed as a quantization procedure, is invertible in the following sense: to every continuous linear operator $\widehat{A} : \mathcal{S}(\mathbb{R}^n) \longrightarrow \mathcal{S}'(\mathbb{R}^n)$ corresponds a unique distribution $a \in \mathcal{S}'(\mathbb{R}^{2n})$ such that $\widehat{A}_{\mathrm{W}} = \mathrm{Op}_{\mathrm{W}}(a)$; conversely given a symbol $a \in \mathcal{S}'(\mathbb{R}^{2n})$ we know how to construct the operator $\mathrm{Op}_{\mathrm{W}}(a)$. Thus,

$$\mathrm{Op}_{\mathrm{W}} : \mathcal{S}'(\mathbb{R}^{2n}) \longrightarrow \mathcal{L}(\mathcal{S}(\mathbb{R}^n), \mathcal{S}'(\mathbb{R}^n))$$

is a bijection (in fact a linear isomorphism). We are going to see that the Born–Jordan quantization

$$\mathrm{Op}_{\mathrm{BJ}} : \mathcal{S}'(\mathbb{R}^{2n}) \longrightarrow \mathcal{L}(\mathcal{S}(\mathbb{R}^n), \mathcal{S}'(\mathbb{R}^n))$$

is surjective, but not injective: there are non-zero symbols $a \in \mathcal{S}'(\mathbb{R}^{2n})$ which are mapped to the zero operator. The physical meaning of this particularity is not yet fully understood.

7.4.1 A first non-injectivity result

We prove here a non-uniqueness result; the general question of invertibility, which is much more subtle, will be addressed below.

Proposition 88. *Let $a \in \mathcal{S}'(\mathbb{R}^{2n})$ and $a_{z_0}(z) = e^{-i\sigma(z,z_0)/\hbar}$. We have*

$$\operatorname{Op}_{\mathrm{BJ}}(a) = \operatorname{Op}_{\mathrm{BJ}}(a + a_{z_0}) \tag{7.16}$$

for all $z_0 = (x_0, p_0)$ such that $F_\sigma\theta_{\mathrm{BJ}}(z_0) = 0$, that is $p_0x_0 = 2N\pi\hbar$ $(N \in \mathbb{Z})$, $N \neq 0$.

Proof. By linearity it suffices to prove that $\operatorname{Op}_{\mathrm{BJ}}(a_{z_0}) = 0$ if $p_0x_0 = 2N\pi\hbar$. We have, by formula (7.9),

$$\operatorname{Op}_{\mathrm{BJ}}(a_{z_0})\psi = \left(\frac{1}{2\pi\hbar}\right)^n \int_{\mathbb{R}^{2n}} F_\sigma a_{z_0}(z)\widehat{D}_{\mathrm{BJ}}(z)\psi dz. \tag{7.17}$$

Now,

$$F_\sigma a_{z_0}(z) = \left(\frac{1}{2\pi\hbar}\right)^n \int_{\mathbb{R}^{2n}} e^{-\frac{i}{\hbar}\sigma(z-z_0,z')}dz' = (2\pi\hbar)^n\delta(z-z_0)$$

and hence

$$\begin{aligned}\operatorname{Op}_{\mathrm{BJ}}(a_{z_0}) &= \int_{\mathbb{R}^{2n}} \delta(z-z_0)\operatorname{sinc}\left(\frac{px}{2\hbar}\right)\widehat{D}(z)dz \\ &= \int_{\mathbb{R}^{2n}} \delta(z-z_0)\operatorname{sinc}\left(\frac{p_0x_0}{2\hbar}\right)dz\end{aligned}$$

so that $\operatorname{Op}_{\mathrm{BJ}}(a_{z_0}) = 0$ if $p_0x_0 \in 2\pi\hbar\mathbb{Z}$ with $p_0x_0 \neq 0$. □

Observe that the set Σ of all z such that $\chi_{\mathrm{BJ}}(z) = 0$ consists of a family of concentric $2n-1$ dimensional sheets in phase space (when $n = 1$ they are just hyperbolas in the phase plane). The distance of the set Σ to the origin is easily calculated and one gets

$$\operatorname{dist}(0, \Sigma) = \sqrt{4\pi\hbar}. \tag{7.18}$$

It follows from (7.16) that we have $\operatorname{Op}_{\mathrm{BJ}}(a) = 0$ for all symbols of the type

$$a(z) = \sum_{z_0\in\Lambda} \lambda(z_0)e^{-i\sigma(z,z_0)/\hbar}$$

where Λ is any finite lattice of points $z_0 = (x_0, p_0)$ in $\mathbb{R}^{2n}$ such that $p_0 x_0 / 2\pi\hbar \in \mathbb{Z}$ and $\lambda : \Lambda \longrightarrow \mathbb{C}$.

However, the Born–Jordan quantization of monomials is invertible:

Assuming that $n = 1$ we denote by $\mathbb{C}[x, p]$ is the polynomial ring generated by the real variables x and p, and by $\mathbb{C}[\widehat{x}, \widehat{p}]$ the algebra generated by $\widehat{x}$ and $\widehat{p}$. Then:

Proposition 89. *The quantization* $\mathrm{Op}_{\mathrm{BJ}}$ *is an isomorphism of vector spaces*

$$\mathrm{Op}_{\mathrm{BJ}} : \mathbb{C}[x, p] \longrightarrow \mathbb{C}[\widehat{x}, \widehat{p}].$$

We omit the (quite technical) proof here and refer to the end of the chapter for references.

7.4.2 A general result

Let us recall a few basic concepts from distribution theory and the theory of complex functions. The Fourier transform of a compactly supported function (or distribution) is an entire analytic function (Paley–Wiener theorem). We denote by $B^{2n}(r)$ the closed ball in $\mathbb{R}^{2n}$ centered at the origin and with radius r, and by $\mathcal{E}'(\mathbb{R}^{2n})$ the space of compactly supported distributions on $\mathbb{R}^{2n}$; we have inclusion $\mathcal{E}'(\mathbb{R}^{2n}) \subset \mathcal{S}'(\mathbb{R}^{2n})$. For $r \geq 0$ we denote by $A_r(2n)$ the subspace of $\mathcal{S}'(\mathbb{R}^{2n})$ consisting of all tempered distributions a whose symplectic Fourier transform $F_\sigma a = a_\sigma$ has support $\operatorname{supp}(a_\sigma) \subset B^{2n}(r)$. Equivalently, a satisfies an estimate

$$|a(\zeta)| \leq C(1 + |\zeta|)^N e^{\frac{r}{\hbar}|\operatorname{Im}\zeta|} \tag{7.19}$$

for some constants $C > 0$, $N > 0$. Obviously $A_0(2) = \mathbb{C}[x, p]$, the space of polynomials in the real variables x and p. More generally, $A_0(2n)$ is the space of polynomials in the variables $x_1, \ldots, x_n$ and $p_1, \ldots, p_n$.

The following inversion result generalizes Proposition 89:

Proposition 90. *The linear mapping* $\mathcal{S}'(\mathbb{R}^{2n}) \longrightarrow \mathcal{S}'(\mathbb{R}^{2n})$ *defined by*

$$a \longmapsto \left(\frac{1}{2\pi\hbar}\right)^n a * \theta_{\mathrm{BJ}} \tag{7.20}$$

restricts to an automorphism of $A_r(2n)$ *if and only if*

$$0 \leq r < \sqrt{4\pi\hbar}. \tag{7.21}$$

That is, the equation $b_\sigma = a_\sigma F_\sigma \theta_{\mathrm{BJ}}$ *admits, for every* $b \in A_r(2n)$, *a unique solution* $a \in A_r(2n)$ *if and only if condition (7.21) holds.*

Proof. Let us first prove the sufficiency of condition (7.21). Assume that $0 \leq r < \sqrt{4\pi\hbar}$; it follows from the equality (7.18) that the ball $B^{2n}(r)$ does not contain any zero of $F_\sigma\theta_{\mathrm{BJ}}$ hence the equation $b_\sigma = a_\sigma F_\sigma\theta_{\mathrm{BJ}}$ admits the solution $a_\sigma = b_\sigma/F_\sigma\theta_{\mathrm{BJ}}$ for every $b \in A_r(2n)$, and it is clear that $a_\sigma \in \mathcal{S}'(\mathbb{R}^{2n})$. Since $\mathrm{supp}(b_\sigma) \subset B^{2n}(r)$ we also have $\mathrm{supp}(a_\sigma) \subset B^{2n}(r)$, hence $a \in A_r(2n)$. Condition (7.21) is also necessary: assume in fact that $r \geq \sqrt{4\pi\hbar}$ and choose $z_0 = (x_0, p_0)$ such that $F_\sigma\theta_{\mathrm{BJ}} = 0$. Then $\widehat{D}_{\mathrm{BJ}}(z_0) = 0$ and $\mathrm{supp}(b_\sigma)$ thus contains the points z_0; hence the mapping (7.20) is not injective. □

As for the announced surjectivity, we have th following result:

Proposition 91. *The mapping* $a \longmapsto (2\pi\hbar)^{-n} a * \theta_{\mathrm{BJ}}$ *is a linear surjection* $\mathcal{S}'(\mathbb{R}^{2n}) \longrightarrow \mathcal{S}'(\mathbb{R}^{2n})$*: for every* $b \in \mathcal{S}'(\mathbb{R}^{2n})$ *there exists (a non-unique)* $a \in \mathcal{S}'(\mathbb{R}^{2n})$ *such that*

$$b = \left(\frac{1}{2\pi\hbar}\right)^n a * \theta_{\mathrm{BJ}}. \tag{7.22}$$

Its proof is highly non-trivial in spite of its apparent simplicity. The difficulty comes from the fact that we are confronted to a division problem for distributions. See the Comments and References section.

7.5 Comments and references

Born–Jordan quantization was proposed in the papers [11, 12] by Pascual Jordan, Max Born, and Werner Heisenberg. The idea of averaging τ-Wigner transforms (which is the key to Born–Jordan quantization) is originally due to Paolo Boggiatto and his collaborators [8–10]. Proposition 82 on the orderings of monomials is due Domingo and Galapon [22].

The section addressing the (non-) invertibility and surjectivity of Born–Jordan quantization originates in joint work with Elena Cordero and Fabio Nicola (in particular the proof of Propositions 90 and 91); see [16]. Also see our monograph [33] for details and comments. The proof of Proposition 87 can be found in our joint work [38] with Fabio Nicola.

The debate is still open whether Born–Jordan quantization is the "right" quantization procedure as opposed to the Weyl transformation is open. The two theories are hard to distinguish, because they both quantize "physical Hamiltonians" of the type

$$H(x, p, t) = \frac{1}{2m}(p - A(x, t))^2 + V(x, t)$$

in the same way, and lead to the operator

$$\widehat{H} = \frac{1}{2m}(-i\hbar\partial_x - A(x, t))^2 + V(x, t)$$

already "guessed" by Erwin Schrödinger. One can prove (see [38]) that the Born–Jordan quantization satisfies the Dirac prescription (1.25) is satisfied for all Hamiltonians of the type above.

8 Metaplectic operators

8.1 The symplectic group Sp(n)

The symplectic group Sp(n) is generated by the set of all free symplectic matrices

$$S = \begin{pmatrix} A & B \\ C & D \end{pmatrix}, \quad \det B \neq 0\,; \tag{8.1}$$

symplectic matrices of this type are called *free symplectic matrices*. The notion depends on the choice of symplectic basis and is therefore not intrinsic. Condition (8.1) is actually equivalent to the transversality condition

$$S(0 \times \mathbb{R}^n) \cap (0 \times \mathbb{R}^n) = 0.$$

We have the following remarkable result (proof omitted):

Proposition 92. *Every $S \in \mathrm{Sp}(n)$ can be written (non-uniquely) as the product of exactly two free symplectic matrices.*

Let L and P be two real $n \times n$ matrices, P symmetric and L invertible. We set

$$V_{-P} = \begin{pmatrix} I & P \\ 0 & I \end{pmatrix}, \quad M_L = \begin{pmatrix} L^{-1} & 0 \\ 0 & L^T \end{pmatrix}.$$

These matrices are trivially symplectic, and one verifies by a straightforward calculation that

$$S = V_{-DB^{-1}} M_{B^{-1}} J V_{-B^{-1}A}. \tag{8.2}$$

It follows from Proposition 92 that:

Proposition 93. *The symplectic group* Sp(n) *is generated by the set*

$$G = \{V_{-P}, M_L, J : P = P^T, \det L \neq 0\}.$$

It is immediately verifiable that these generators satisfy the relationship

$$V_{-P} V_{-P'} = V_{-(P+P')}, \quad M_L M_{L'} = M_{L'L} \tag{8.3}$$

$$M_L V_{-P} = V_{-L^T P L} M_L, \quad V_{-P} M_L = M_L V_{-(L^{-1})^T P L^{-1}} \tag{8.4}$$

and

$$(V_{-P} M_L)^{-1} = V_{-(L^{-1})^T P L^{-1}} M_{L^{-1}}. \tag{8.5}$$

An immediate consequence of this is that the determinant of a symplectic matrix always is equal to +1 (while the defining definition $S^T J S = J$ of a symplectic matrix only allows to conclude that $\det S = \pm 1$).

https://doi.org/10.1515/9783110722772-008

To each free symplectic matrix (8.1), we associate a homogeneous quadratic polynomial in the variables (x, x'):

$$W(x,x') = \frac{1}{2}DB^{-1}x^2 - B^{-1}x \cdot x' + \frac{1}{2}B^{-1}Ax'^2.$$

This polynomial is called the generating function of S, and we have

$$(x,p) = S(x',p') \Longleftrightarrow \begin{cases} p = \partial_x W(x,x') \\ p' = -\partial_{x'} W(x,x'). \end{cases}$$

Conversely, to every quadratic polynomial of the type

$$\begin{aligned} W(x,x') &= \frac{1}{2}Px^2 - Lx \cdot x' + \frac{1}{2}Qx'^2 \\ P = P^T, &\quad Q = Q^T, \quad \det L \neq 0, \end{aligned} \tag{8.6}$$

we can associate a free symplectic matrix, namely

$$S_W = \begin{pmatrix} L^{-1}Q & L^{-1} \\ PL^{-1}Q - L^T & PL^{-1} \end{pmatrix}. \tag{8.7}$$

The latter can be factorized as

$$S_W = V_{-P}M_L J V_Q. \tag{8.8}$$

8.2 The metaplectic representation

8.2.1 Fourier integral operators with quadratic phases

The metaplectic group can be defined in several ways. Our approach highlights the notion of generating the function of a symplectic matrix that allows defining the metaplectic group in terms of a class of Fourier integral operators with quadratic phase.

To every free symplectic matrix S_W, we associate an operator $\widehat{S}_{W,m}$ by setting, for $\psi \in \mathcal{S}(\mathbb{R}^n)$,

$$\widehat{S}_{W,m}\psi(x) = \left(\frac{1}{2\pi i\hbar}\right)^{n/2} \Delta(W) \int_{\mathbb{R}^n} e^{\frac{i}{\hbar}W(x,x')}\psi(x')dx'\,; \tag{8.9}$$

here $\arg i = \pi/2$ and the prefactor $\Delta(W)$ is defined by

$$\Delta(W) = i^m \sqrt{|\det L|}\,; \tag{8.10}$$

the integer m ("Maslov index") corresponds to a choice of $\arg \det L$:

$$m\pi \equiv \arg \det L \mod 2\pi. \tag{8.11}$$

(Thus to each S_W correspond *two* operators: $\widehat{S}_{W,m}$ and $\widehat{S}_{W,m+2} = -\widehat{S}_{W,m}$). We will call the operator $\widehat{S}_{W,m}$ the "quadratic Fourier transform" associated with the free symplectic matrix S_W.

The generating function of the standard symplectic J is $W(x,x') = -x \cdot x'$. We denote by

$$\widehat{J}\psi(x) = \left(\frac{1}{2\pi i}\right)^{n/2} \int_{\mathbb{R}^n} e^{-\frac{i}{\hbar}x\cdot x'}\psi(x')dx' = i^{-n/2}F\psi(x) \tag{8.12}$$

the operator corresponding to the choice $m = 0$ for $L = I_{n\times n}$. For $\psi \in \mathcal{S}(\mathbb{R}^n)$; F is the usual unitary Fourier transform on $\mathbb{R}^n$.

Let us define operators $\widehat{V}_{-P}$ and $\widehat{M}_{L,m}$ by

$$\widehat{V}_{-P}\psi(x) = e^{\frac{i}{2}Px\cdot x}\psi(x), \quad \widehat{M}_{L,m}\psi(x) = i^m\sqrt{|\det L|}\psi(Lx). \tag{8.13}$$

These are obviously unitary operators on $L2(\mathbb{R}^n)$. Let W be the quadratic form (8.6). We have the factorization

$$\widehat{S}_{W,m} = \widehat{V}_{-P}\widehat{M}_{L,m}\widehat{J}\widehat{V}_{-Q}\,. \tag{8.14}$$

This immediately follows from the definition of $\widehat{J}$, noting that

$$\widehat{M}_{L,m}\widehat{J}\psi(x) = \left(\frac{1}{2\pi i}\right)^{n/2} i^m\sqrt{|\det L|}\int_{\mathbb{R}^n} e^{-iLx\cdot x'}\psi(x')dx'.$$

Proposition 94. *The operators $\widehat{S}_{W,m}$ extend to unitary operators $L^2(\mathbb{R}^n) \longrightarrow L^2(\mathbb{R}^n)$, and the inverse of $\widehat{S}_{W,m}$ is given by*

$$\widehat{S}_{W,m}^{-1} = \widehat{S}_{W^*,m^*}, \quad W^*(x,x') = -W(x',x), \quad m^* = n-m. \tag{8.15}$$

Proof. We obviously have

$$(\widehat{V}_{-P})^{-1} = \widehat{V}_P \quad \text{and} \quad (\widehat{M}_{L,m})^{-1} = \widehat{M}_{L^{-1},-m},$$

and $\widehat{J}^{-1}$ is given by the Fourier inversion formula:

$$\widehat{J}^{-1}\psi(x) = \left(\frac{i}{2\pi\hbar}\right)^{n/2}\int_{\mathbb{R}^n} e^{\frac{i}{\hbar}x\cdot x'}\psi(x')dx'.$$

In view of the identity (8.14), we have

$$\widehat{S}_{W,m}^{-1} = \widehat{V}_Q\widehat{J}^{-1}\widehat{M}_{L^{-1},-m}\widehat{V}_P,$$

and the inversion formulas (8.15) follow from the fact that

$$\begin{aligned}\widehat{J}^{-1}\widehat{M}_{L^{-1},-m}\psi(x) &= \left(\frac{i}{2\pi}\right)^{n/2} i^{-m}\sqrt{|\det L^{-1}|}\int_{\mathbb{R}^n} e^{\frac{i}{\hbar}x\cdot x'}\psi(L^{-1}x')dx' \\ &= \left(\frac{1}{2\pi i}\right)^{n/2} i^{-m+n}\sqrt{|\det L|}\int_{\mathbb{R}^n} e^{\frac{i}{\hbar}L^T x\cdot x'}\psi(x')dx' \\ &= \widehat{M}_{-L^T,n-m}\widehat{J}\psi(x). \end{aligned}$$

□

8.2.2 Definition and properties of Mp(n)

As a consequence of the proposition that the operators $\widehat{S}_{W,m}$ form a subset of the group $\mathcal{U}(L^2(\mathbb{R}^n))$ of unitary operators acting on $L^2(\mathbb{R}^n)$, this subset is closed under the operation of inversion. This motivates the following definition:

Definition 95. The subgroup of $\mathcal{U}(L^2(\mathbb{R}^n))$ generated by the quadratic Fourier transforms $\widehat{S}_{W,m}$ is called the "metaplectic group" and is denoted by $\mathrm{Mp}(n)$. The elements of $\mathrm{Mp}(n)$ are called "metaplectic operators".

The following result considerably simplifies many arguments; it is the metaplectic analogue of Proposition 92 which states that every symplectic matrix can be written as the product of two free symplectic matrices.

Proposition 96. (i) *Every $\widehat{S} \in \mathrm{Mp}(n)$ can be written as a product of exactly two quadratic Fourier transforms: $\widehat{S} = \widehat{S}_{W,m}\widehat{S}_{W',m'}$.* (ii) *The metaplectic group $\mathrm{Mp}(n)$ is generated by the operators $\widehat{V}_{-P}$, $\widehat{M}_{L,m}$, and $\widehat{J}$.*

Proof. (i) The result follows from the existence of a natural projection $\mathrm{Mp}(n) \longrightarrow \mathrm{Sp}(n)$, which will be established in next section, and the fact that every $S \in \mathrm{Sp}(n)$ can be written as a product $S_W S_{W'}$. (See the Comments at the end of the chapter). (ii) It follows from the definition of $\mathrm{Mp}(n)$, together with the fact that each $\widehat{S}_{W,m}$ is a product $\widehat{V}_{-P}\widehat{M}_{L,m}\widehat{J}\widehat{V}_{-Q}$ (formula (8.14)). (ii) It immediately follows in view of the factorization formula (8.14). □

The factorization $\widehat{S} = \widehat{S}_{W,m}\widehat{S}_{W',m'}$, however, is not unique. For instance, the identity operator can be written $\widehat{S}_{W,m}\widehat{S}_{W^*,m^*}$ for every generating function W.

8.3 The projection π^{Mp}

8.3.1 Definition of the covering projection

We are going to show that $\mathrm{Mp}(n)$ is a double covering group of $\mathrm{Sp}(n)$ by defining a covering mapping $\pi^{\mathrm{Mp}} : \mathrm{Mp}(n) \longrightarrow \mathrm{Sp}(n)$.

Proposition 97. *The mapping* $\widehat{S}_{W,m} \longmapsto S_W$ *extends into a unique group epimorphism*

$$\pi^{\mathrm{Mp}} : \mathrm{Mp}(n) \longrightarrow \mathrm{Sp}(n),$$

with kernel $\ker(\pi^{\mathrm{Mp}}) = \{-I, +I\}$ *and satisfying*

$$\pi^{\mathrm{Mp}}(\widehat{J}) = J, \quad \pi^{\mathrm{Mp}}(\widehat{M}_{L,m}) = M_L, \quad \pi^{\mathrm{Mp}}(\widehat{V}_P) = V_P. \tag{8.16}$$

Proof. The construction of π^{Mp} will be given later; let us prove the formulas (8.16) assuming that $\pi^{\mathrm{Mp}}(\widehat{S}\widehat{S}') = \pi^{\mathrm{Mp}}(\widehat{S})\pi^{\mathrm{Mp}}(\widehat{S}\widehat{S}')$. The formula $\pi^{\mathrm{Mp}}(\widehat{J}) = J$ is obvious since $\widehat{J}$ is a quadratic Fourier transform with $W(x,x') = -x \cdot x'$. We have $\widehat{M}_{L,m} = \widehat{J}^{-1}(\widehat{J}\widehat{M}_{L,m})$ and $\widehat{J}\widehat{M}_{L,m} = \widehat{S}_{W,m}$ with $W(x,x') = -(L^T)^{-1}x \cdot x'$. Since $\pi^{\mathrm{Mp}}(\widehat{J}^{-1}) = J^{-1} = -J$, it follows that

$$\pi^{\mathrm{Mp}}(\widehat{M}_{L,m}) = \pi^{\mathrm{Mp}}(\widehat{J}^{-1})\pi^{\mathrm{Mp}}(\widehat{J}\widehat{M}_{L,m}) = M_L.$$

The formula $\pi^{\mathrm{Mp}}(\widehat{V}_P) = V_P$ is proven using a similar argument. □

8.3.2 Construction of π^{Mp}

Let us now give an explicit construction of the covering projection π^{Mp}. For notational simplicity, we take $\hbar = 1$.

Let us denote the elements of the dual $(\mathbb{R}^{2n})^*$ of $\mathbb{R}^{2n}$ by the Latin letters a, b, c, etc. Thus $a(z) = a(x,p)$ is the value of the linear form a at the point $z = (x,p)$. To every $a \in (\mathbb{R}^{2n})^*$, we associate a first-order linear partial differential operator A obtained by replacing formally p in $a(x,p)$ by D_x:

$$A = a(x, D_x), \quad D_x = -i\partial_x;$$

thus, if

$$a(x,p) = \alpha \cdot x + \beta \cdot p$$

for $\alpha = (\alpha_1, \dots, \alpha_n)$, $\beta = (\beta_1, \dots, \beta_n)$ in $\mathbb{R}^n$, then

$$A = \alpha \cdot x + \beta \cdot D_x = \alpha \cdot x - i\beta \cdot \partial_x. \tag{8.17}$$

Obviously the sum of two operators of this type is an operator of the same type, and so is the product of such an operator by a scalar. It follows that these operators form a $2n$-dimensional vector space, which we denote by $\mathrm{Diff}^{(1)}(n)$.

The vector spaces $\mathbb{R}^{2n}$, $(\mathbb{R}^{2n})^*$ and $\mathrm{Diff}^{(1)}(n)$ are isomorphic since they all have the same dimension $2n$. The following result explicitly describes three canonical isomorphisms between these spaces:

Lemma 98. (i) *The linear mappings*

$$\varphi_1 : \mathbb{R}^{2n} \longrightarrow (\mathbb{R}^{2n})^*, \quad \varphi_1 : z_0 \longmapsto a$$
$$\varphi_2 : (\mathbb{R}^{2n})^* \longrightarrow \mathrm{Diff}^{(1)}(n), \quad \varphi_2 : a \longmapsto A$$

where a is the unique linear form on $\mathbb{R}^{2n}$ such that $a(z) = \sigma(z, z_0)$ are isomorphisms, hence their compose is φ:

$$\varphi = \varphi_2 \circ \varphi_1 : \mathbb{R}^{2n} \longrightarrow \mathrm{Diff}^{(1)}(n);$$

the latter associates with $z_0 = (x_0, p_0)$ the operator

$$A = \varphi(z_0) = p_0 \cdot x - x_0 \cdot D_x.$$

(ii) *Let $[A, B] = AB - BA$ be the commutator of $A, B \in \mathrm{Diff}^{(1)}(n)$; we have*

$$[\varphi(z_1), \varphi(z_2)] = -i\sigma(z_1, z_2) \tag{8.18}$$

for all $z_1, z_2 \in \mathbb{R}^{2n}$.

Proof. (i) For the vector spaces $\mathbb{R}^{2n}$, $(\mathbb{R}^{2n})^*$, and $\mathrm{Diff}^{(1)}(n)$ having the same dimension, it suffices to show that $\ker(\varphi_1)$ and $\ker(\varphi_2)$ are zero. Now, $\varphi_1(z_0) = 0$ is equivalent to the condition $\sigma(z, z_0) = 0$ for all z, and hence to $z_0 = 0$ since a symplectic form is non-degenerate. If $\varphi_2(a) = 0$, then

$$A\psi = \varphi_2(a)\psi = 0 \quad \text{for all } \psi \in \mathcal{S}(\mathbb{R}^n)$$

which implies $A = 0$ and thus $a = 0$. (ii) Let $z_1 = (x_1, p_1)$, $z_2 = (x_2, p_2)$. We have

$$\varphi(z_1) = p_1 \cdot x - x_1 \cdot D_x, \quad \varphi(z_2) = p_2 \cdot x - x_2 \cdot D_x$$

and hence

$$[\varphi(z_1), \varphi(z_2)] = i(x_1 \cdot p_2 - x_2 \cdot p_1),$$

which is precisely the commutation formula (8.18). □

We are next going to show that the metaplectic group $\mathrm{Mp}(n)$ acts by conjugation on $\mathrm{Diff}^{(1)}(n)$.

Lemma 99. *For $z_0 = (x_0, p_0) \in \mathbb{R}^{2n}$, define $\widehat{A} \in \mathrm{Diff}^{(1)}(n)$ by*

$$\widehat{A} = \varphi(z_0) = p_0 \cdot x - x_0 \cdot D_x.$$

(i) *We have:*

$$\widehat{J}\widehat{A}\widehat{J}^{-1} = -x_0 \cdot x - p_0 \cdot D_x = \varphi(Jz_0) \tag{8.19}$$

$$\widehat{M}_{L,m}\widehat{A}(\widehat{M}_{L,m})^{-1} = L^T p_0 \cdot x - L^{-1}x_0 \cdot D_x = \varphi(M_L z_0) \tag{8.20}$$

$$\widehat{V}_P\widehat{A}(\widehat{V}_P)^{-1} = (p_0 + Px_0) \cdot x - x_0 \cdot D_x = \varphi(V_P z_0). \tag{8.21}$$

(ii) *If* $\widehat{A} \in \mathrm{Diff}^{(1)}(n)$ *and* $\widehat{S} \in \mathrm{Mp}(n)$, *then* $\widehat{S}\widehat{A}\widehat{S}^{-1} \in \mathrm{Diff}^{(1)}(n)$. (iii) *For every* $\widehat{S} \in \mathrm{Mp}(n)$, *the mapping*

$$\Phi_{\widehat{S}} : \mathrm{Diff}^{(1)}(n) \longrightarrow \mathrm{Diff}^{(1)}(n), \quad \widehat{A} \longmapsto \widehat{S}\widehat{A}\widehat{S}^{-1}$$

is a vector space automorphism.

Proof. (i) Using the properties of the Fourier transform, it is immediately verifiable that:

$$(x_0 \cdot D_x)\psi = \widehat{J}^{-1}(x_0 \cdot x)\widehat{J}\psi$$
$$(p_0 \cdot x)\psi = -\widehat{J}^{-1}(p_0 \cdot D_x)\widehat{J}\psi$$

for $\psi \in \mathcal{S}(\mathbb{R}^n)$, hence (8.19). To prove (8.20), it suffices to remark that

$$\widehat{M}_{L,m}(p_0 \cdot x)(\widehat{M}_{L,m})^{-1}\psi(x) = (p_0 \cdot Lx)\psi(x)$$

and

$$\widehat{M}_{L,m}(x_0 \cdot D_x)(\widehat{M}_{L,m})^{-1}\psi(x) = x_0(L^{-1})^T D_x\psi(x).$$

Let us prove formula (8.21). Recalling that by definition

$$\widehat{V}_{-P}\psi(x) = e^{\frac{i}{2}Px\cdot x}\psi(x),$$

we have, since P is symmetric,

$$(x_0 \cdot D_x)\widehat{V}_{-P}\psi(x) = \widehat{V}_{-P}(Px_0 \cdot x)\psi(x) + (p_0 \cdot D_x)\psi(x),$$

and hence

$$\widehat{V}_P A(\widehat{V}_{-P}\psi)(x) = ([p_0 + Px_0) \cdot x]\psi(x) - (x_0 \cdot D_x)\psi(x)$$

which is (8.21). (ii) This property immediately follows since $\widehat{S}$ is a product of operators $\widehat{J}, \widehat{M}_{L,m}, \widehat{V}_P$. (iii) The mapping $\Phi_{\widehat{S}}$ is trivially a linear mapping $\mathrm{Diff}^{(1)}(n) \longrightarrow \mathrm{Diff}^{(1)}(n)$. If $B = \widehat{S}\widehat{A}\widehat{S}^{-1} \in \mathrm{Diff}^{(1)}(n)$, then we have also $\widehat{A} = \widehat{S}^{-1}\widehat{B}\widehat{S} \in \mathrm{Diff}^{(1)}(n)$ since $\widehat{A} = \widehat{S}^{-1}\widehat{B}(\widehat{S}^{-1})^{-1}$. It follows that $\Phi_{\widehat{S}}$ is surjective and hence bijective. □

Since the operators $\widehat{J}, \widehat{M}_{L,m}, \widehat{V}_P$ generate $\mathrm{Mp}(n)$, the Lemma just stated shows that, for every $\widehat{S} \in \mathrm{Mp}(n)$, there exists a linear automorphism S of $\mathbb{R}^{2n}$ such that $\Phi_{\widehat{S}}(\widehat{A}) = \widehat{a \circ S}$ that is

$$\Phi_{\widehat{S}}(\varphi(z_0)) = \varphi(Sz_0). \tag{8.22}$$

Let us show that the automorphism S preserves the symplectic form. For $z, z' \in \mathbb{R}^{2n}$, we have, in view of the commutation formula (8.18),

$$\begin{aligned}\sigma(Sz, Sz') &= i[\varphi(Sz), \varphi(Sz')] \\ &= i[\Phi_{\widehat{S}}\varphi(z), \Phi_{\widehat{S}}\varphi(z')] \\ &= i[\widehat{S}\varphi(z)\widehat{S}^{-1}, \widehat{S}\varphi(z')\widehat{S}^{-1}] \\ &= i\widehat{S}[\varphi(z), \varphi(z')]\widehat{S}^{-1} \\ &= \sigma(z, z'),\end{aligned}$$

hence $S \in \mathrm{Sp}(n)$ as claimed.

We are now able to describe explicitly the natural projection of $\mathrm{Mp}(n)$ onto $\mathrm{Sp}(n)$.

Definition 100. The covering projection $\pi^{\mathrm{Mp}} : \mathrm{Mp}(n) \longrightarrow \mathrm{Sp}(n)$ is the mapping π^{Mp}, which to $\widehat{S} \in \mathrm{Mp}(n)$ associates the element $\pi^{\mathrm{Mp}}(\widehat{S}) = S \in \mathrm{Sp}(n)$ defined by (8.22), that is

$$S = \varphi^{-1}\Phi_{\widehat{S}}\varphi. \tag{8.23}$$

That the mapping π^{Mp} indeed is a covering mapping follows from:

Proposition 101. (i) *The mapping π^{Mp} is a continuous group epimorphism of* $\mathrm{Mp}(n) \longrightarrow \mathrm{Sp}(n)$, *such that:*

$$\pi^{\mathrm{Mp}}(\widehat{J}) = J, \quad \pi^{\mathrm{Mp}}(\widehat{M}_{L,m}) = M_L, \quad \pi^{\mathrm{Mp}}(\widehat{V}_P) = V_P, \tag{8.24}$$

and hence

$$\pi^{\mathrm{Mp}}(\widehat{S}_{W,m}) = S_W. \tag{8.25}$$

(ii) *We have* $\ker(\pi^{\mathrm{Mp}}) = \{-I, +I\}$; *hence* $\pi^{\mathrm{Mp}} : \mathrm{Mp}(n) \longrightarrow \mathrm{Sp}(n)$ *is a two-fold covering map.*

Proof. (i) Let us first show that π^{Mp} is a group homomorphism. In view of the obvious identity $\Phi_{\widehat{S}}\Phi_{\widehat{S}'} = \Phi_{\widehat{S}\widehat{S}'}$ we have

$$\begin{aligned}\pi^{\mathrm{Mp}}(\widehat{S}\widehat{S}') &= \varphi^{-1}\Phi_{\widehat{S}\widehat{S}'}\varphi \\ &= (\varphi^{-1}\Phi_{\widehat{S}}\varphi)(\varphi^{-1}\Phi_{\widehat{S}'}\varphi) \\ &= \pi^{\mathrm{Mp}}(\widehat{S})\pi^{\mathrm{Mp}}(\widehat{S}').\end{aligned}$$

Let us next prove that π^{Mp} is surjective. Recall that the matrices J,

$$M_L = \begin{pmatrix} L^{-1} & 0 \\ 0 & L^T \end{pmatrix} \quad \text{and} \quad V_{-P} = \begin{pmatrix} I & 0 \\ P & I \end{pmatrix}$$

generate Sp(n) when L and P range over, respectively, the invertible and symmetric real matrices of order n. It is thus sufficient to show that formulae (8.24) hold. Now, using (8.19), (8.20), and (8.21), we have

$$\varphi\Phi_{\widehat{J}}\varphi^{-1} = J, \quad \varphi\Phi_{\widehat{M}_{L,m}}\varphi^{-1} = M_L, \quad \varphi\Phi_{\widehat{V}_{-P}}\varphi^{-1} = V_{-P},$$

hence (8.16). Formula (8.25) follows since every quadratic Fourier transform $\widehat{S}_{W,m}$ can be factorized as

$$\widehat{S}_{W,m} = \widehat{V}_{-P}\widehat{M}_{L,m}\widehat{J}\widehat{V}_{-Q}$$

in view of Proposition 94. To establish the continuity of the mapping π^{Mp}, we first remark that the isomorphism $\varphi : \mathbb{R}^{2n} \longrightarrow \mathrm{Diff}^{(1)}(n)$ defined in Lemma 98 is trivially continuous, and so is its inverse. Since $\Phi_{\widehat{S}\widehat{S}'} = \Phi_{\widehat{S}}\Phi_{\widehat{S}'}$ it suffices to show that for every $\widehat{A} \in \mathrm{Diff}^{(1)}(n)$, $\psi_S(\widehat{A})$ has $\widehat{A}$ as limit when $\widehat{S} \to \widehat{I}$ in Mp(n). Now, Mp(n) is a group of continuous automorphisms of $\mathcal{S}(\mathbb{R}^n)$ hence, when $\widehat{S} \to \widehat{I}$ then $\widehat{S}^{-1}\psi \to \psi$ for every $\psi \in \mathcal{S}(\mathbb{R}^n)$, that is $\widehat{A}\widehat{S}^{-1}\psi \to \widehat{A}\psi$ and also $\widehat{S}\widehat{A}\widehat{S}^{-1}\psi \to \psi$. (ii) Suppose that $\varphi^{-1}\Phi_{\widehat{S}}\varphi = \widehat{I}$. Then $\widehat{S}\widehat{A}\widehat{S}^{-1} = \widehat{A}$ for every $\widehat{A} \in \mathrm{Diff}^{(1)}(n)$, and this is only possible if $\widehat{S}$ is multiplication by a constant c with $|c| = 1$, and thus $\ker(\pi^{\mathrm{Mp}}) \subset S^1$. In view of Lemma 96, we have $\widehat{S} = \widehat{S}_{W,m}\widehat{S}_{W',m'}$ for some choice of (W, m) and (W', m'), hence the condition $\widehat{S} \in \ker(\pi^{\mathrm{Mp}})$ is equivalent to

$$\widehat{S}_{W',m'} = c(\widehat{S}_{W,m})^{-1} = c\widehat{S}_{W^*,m^*}$$

which is only possible if $c = \pm 1$, hence $\widehat{S} = \pm\widehat{I}$ as claimed. □

8.4 Comments and references

See Jean Leray [54], Chap. 1, or de Gosson [29]. According to Leray, the approach to the metaplectic group using the quadratic Fourier integral operators is due to Vladimir Buslaev. An excellent treatment of the metaplectic group, much in the spirit of Leray, is given by Hans Reiter [63].

For other explicit formulas giving the action of the metaplectic group on Gaussians, see Folland [28].

9 The property of symplectic covariance

The property of symplectic covariance is of fundamental importance in harmonic analysis. This property is characteristic of Weyl quantization and allows considerably simplifying many proofs and the statement of many properties.

9.1 Symplectic covariance of the cross-Wigner transform

9.1.1 The Heisenberg displacement operators

The phase-space translation operators $T(z_0)$ satisfy the intertwining formula $ST(z_0)S^{-1} = T(Sz_0)$ for every $S \in \mathrm{Sp}(n)$. It is therefore perhaps not so surprising that we have a similar formula for the Heisenberg displacement operators. We are going to prove a symplectic covariance formula for these operators; this result is fundamental because it implies the symplectic covariance of the Wigner transform ad Weyl calculus.

Proposition 102. *Let* $\widehat{S} \in \mathrm{Mp}(n)$ *and* $S = \pi^{\mathrm{Mp}}(\widehat{S})$. *We have*

$$\widehat{S}\widehat{D}(z_0)\widehat{S}^{-1} = \widehat{D}(Sz_0) \tag{9.1}$$

for every $z_0 \in \mathbb{R}^{2n}$.

Proof. It is sufficient to assume that $\widehat{S}$ is a quadratic Fourier transform $\widehat{S}_{W,m}$ since every $\widehat{S} \in \mathrm{Mp}(n)$ is a product of two such operators. Suppose indeed that we have shown that

$$\widehat{D}(S_W z_0) = \widehat{S}_{W,m}\widehat{D}(z_0)\widehat{S}_{W,m}^{-1}; \tag{9.2}$$

writing an arbitrary element S of $\mathrm{Mp}(n)$ as a product $S_{W,m}S_{W',m'}$ we have

$$\begin{aligned}
\widehat{D}(Sz_0) &= \widehat{S}_{W,m}(\widehat{S}_{W',m'}\widehat{D}(z_0)\widehat{S}_{W',m'}^{-1})\widehat{S}_{W,m}^{-1} \\
&= \widehat{S}_{W,m}\widehat{D}(S_{W'}z_0)\widehat{S}_{W,m}^{-1} \\
&= \widehat{D}(S_W S_{W'} z_0) \\
&= \widehat{S}_{W,m}\widehat{S}_{W',m'}\widehat{D}(z_0)(\widehat{S}_{W,m}\widehat{S}_{W',m'})^{-1} \\
&= \widehat{S}\widehat{D}(z_0)\widehat{S}^{-1}.
\end{aligned}$$

Let us set out to prove that

$$\widehat{D}(z_0)\widehat{S}_{W,m} = \widehat{S}_{W,m}\widehat{D}(S_W^{-1}z_0). \tag{9.3}$$

For $\psi \in \mathcal{S}(\mathbb{R}^n)$ set,

$$g(x) = \widehat{D}(z_0)\widehat{S}_{W,m}\psi(x).$$

https://doi.org/10.1515/9783110722772-009

By definition of a $\widehat{S}_{W,m}$ and $\widehat{D}(z_0)$, we have

$$g(x) = \left(\frac{1}{2\pi i\hbar}\right)^{n/2} \Delta(W) e^{-\frac{1}{2\hbar} p_0 \cdot x_0} \int_{\mathbb{R}^n} e^{\frac{i}{\hbar}(W(x-x_0,x')+p_0\cdot x)} \psi(x') dx'.$$

It is straightforward to verify that the function

$$W_0(x,x') = W(x-x_0,x') + p_0 \cdot x \tag{9.4}$$

is a generating function of the free affine symplectomorphism $T(z_0)S_W$, that is,

$$(x,p) = T(z_0)S_W(x',p') \Longleftrightarrow \begin{cases} p = \partial_x W_0(x,x') \\ p' = -\partial_{x'} W_0(x,x'), \end{cases}$$

hence we have just shown that

$$\widehat{D}(z_0)\widehat{S}_{W,m} = e^{\frac{i}{2\hbar} p_0 \cdot x_0} \widehat{S}_{W_0,m} \tag{9.5}$$

where $\widehat{S}_{W_0,m}$ is one of the metaplectic operators associated to W_0. Let us now set

$$h(x) = \widehat{S}_{W,m}\widehat{D}(S_W^{-1} z_0)\psi(x) \quad \text{and} \quad (x_0', p_0') = \widehat{S}_{W,m}^{-1}(x_0, p_0);$$

we have

$$h(x) = \left(\frac{1}{2\pi i\hbar}\right)^{n/2} \Delta(W) \int_{\mathbb{R}^n} e^{\frac{i}{\hbar} W(x,x')} e^{-\frac{i}{2\hbar} p_0' \cdot x_0'} e^{\frac{i}{\hbar} p_0' \cdot x'} \psi(x' - x_0')\, dx'$$

that is, performing the change of variables $x' \longmapsto x' + x_0'$:

$$h(x) = \left(\frac{1}{2\pi i\hbar}\right)^{n/2} \Delta(W) \int_{\mathbb{R}^n} e^{\frac{i}{\hbar} W(x,x'+x_0')} e^{\frac{i}{2\hbar} p_0' \cdot x_0'} e^{\frac{i}{\hbar} p_0' \cdot x'} \psi(x')\, dx'.$$

We will thus have $h(x) = g(x)$ as claimed, if we show that

$$W(x, x' + x_0') + \frac{1}{2} p_0' \cdot x_0' + p_0' \cdot x' = \text{implies} :_0 (x,x') - \frac{1}{2} p_0 \cdot x_0,$$

that is

$$W(x, x' + x_0') + \frac{1}{2} p_0' \cdot x_0' + p_0' \cdot x' = W(x - x_0, x') + p_0 \cdot x - \frac{1}{2} p_0 \cdot x_0.$$

Replacing x by $x + x_0$, this amounts to proving that

$$W(x + x_0, x' + x_0') + \frac{1}{2} p_0' \cdot x_0' + p_0' \cdot x' = W(x,x') + \frac{1}{2} p_0 \cdot x_0 + p_0 \cdot x,$$

which is straightforward to verify by a direct calculation. □

9.1.2 The Grossmann–Royer reflections

Recall that the Grossmann–Royer reflection operator, which is explicitly given by the formula

$$\widehat{R}(z_0)\psi(x) = e^{\frac{2i}{\hbar}p_0\cdot(x-x_0)}\psi(2x_0 - x),$$

was defined in terms of the displacement operator as

$$\widehat{R}(z_0) = \widehat{D}(z_0)\widehat{R}\widehat{D}(z_0)^{-1}, \quad \widehat{R}\psi(x) = \psi(-x). \tag{9.6}$$

Proposition 102 implies:

Corollary 103. *Let $\widehat{S} \in \mathrm{Mp}(n)$ and $S = \pi^{\mathrm{Mp}}(\widehat{S})$. We have*

$$\widehat{S}\widehat{R}(z_0)\widehat{S}^{-1} = \widehat{R}(Sz_0). \tag{9.7}$$

Proof. Applying the symplectic covariance formula (9.1) to (9.6), we get, since $\widehat{D}(z_0)^{-1} = \widehat{D}(-z_0)$,

$$\begin{aligned}\widehat{S}\widehat{R}(z_0)\widehat{S}^{-1} &= \widehat{S}\widehat{D}(z_0)\widehat{R}\widehat{D}(-z_0)\widehat{S}^{-1}\\ &= (\widehat{S}\widehat{D}(z_0)\widehat{S}^{-1})\widehat{S}\widehat{R}\widehat{S}^{-1}(\widehat{S}\widehat{D}(-z_0)\widehat{S}^{-1})\\ &= \widehat{D}(Sz_0)(\widehat{S}\widehat{R}\widehat{S}^{-1})\widehat{D}(-Sz_0).\end{aligned}$$

Formula (9.7) now follows since $\widehat{S}\widehat{R}\widehat{S}^{-1} = \widehat{R}$ because metaplectic operators commute with reflections. It suffices to prove this claim when $\widehat{S} = \widehat{S}_{W,m}$: for $\psi \in \mathcal{S}(\mathbb{R}^n)$, we have

$$\begin{aligned}\widehat{R}\widehat{S}_{W,m}\psi(x) &= \left(\frac{1}{2\pi i\hbar}\right)^{n/2}\Delta(W)\int_{\mathbb{R}^n} e^{\frac{i}{\hbar}W(-x,x')}\psi(x')\,dx'\\ &= \left(\frac{1}{2\pi i\hbar}\right)^{n/2}\Delta(W)\int_{\mathbb{R}^n} e^{\frac{i}{\hbar}W(-x,-x')}\psi(-x')\,dx'\\ &= \left(\frac{1}{2\pi i\hbar}\right)^{n/2}\Delta(W)\int_{\mathbb{R}^n} e^{\frac{i}{\hbar}W(x,x')}\widehat{R}\psi(x')\,dx'\end{aligned}$$

since $W(-x,-x') = W(x,x')$. □

9.1.3 Symplectic covariance of $W(\psi,\phi)$ and $A(\psi,\phi)$

From the results just stated, we immediately get the desired symplectic covariance result for the cross-Wigner and ambiguity functions:

Proposition 104. *Let* $\widehat{S} \in \mathrm{Mp}(n)$ *and* $S = \pi^{\mathrm{Mp}}(\widehat{S})$ *and* $(\psi, \phi) \in L^2(\mathbb{R}^n) \times L^2(\mathbb{R}^n)$. *We have*

$$W(\widehat{S}\psi, \widehat{S}\phi) = W(\psi, \phi) \circ S^{-1} \quad \text{and} \quad A(\widehat{S}\psi, \widehat{S}\phi) = A(\psi, \phi) \circ S^{-1}.$$

Proof. The cross-Wigner and cross-ambiguity functions are defined in terms of the reflection and displacement operators as (formulas (3.1) and (3.18))

$$W(\psi, \phi) = \left(\frac{1}{\pi\hbar}\right)^n (\widehat{R}(\cdot)\psi|\phi)_{L^2}$$

$$A(\psi, \phi)(z) = \left(\frac{1}{2\pi\hbar}\right)^n (\psi|\widehat{D}(z)\phi)_{L^2}.$$

We have

$$W(\widehat{S}\psi, \widehat{S}\phi) = \left(\frac{1}{\pi\hbar}\right)^n (\widehat{R}(\cdot)\widehat{S}\psi|\widehat{S}\phi)_{L^2},$$

that is, taking the unitarity of $\widehat{S}$ into account and using formula (9.7),

$$\begin{aligned} W(\widehat{S}\psi, \widehat{S}\phi)(z) &= \left(\frac{1}{\pi\hbar}\right)^n (\widehat{S}^{-1}\widehat{R}(z)\widehat{S}\psi|\phi)_{L^2} \\ &= \left(\frac{1}{\pi\hbar}\right)^n (\widehat{R}(S^{-1}z)\psi|\phi)_{L^2} \\ &= W(\psi, \phi)(S^{-1}z). \end{aligned}$$

The symplectic covariance formula for the cross-ambiguity function is proven similarly, using (9.1). □

9.2 Application: the action of Mp(n) on Gaussians

9.2.1 A first result

An immediate consequence of the symplectic covariance of the Wigner transform is the following very useful result that allows to study the action of the metaplectic group on Gaussians up to a phase factor:

Proposition 105. *Let* $\widehat{S} \in \mathrm{Mp}(n)$ *have projection* $S = \begin{pmatrix} A & B \\ C & D \end{pmatrix}$ *on* $\mathrm{Sp}(n)$. *Let*

$$\phi_0(x) = (\pi\hbar)^{-n/4} e^{-|x|^2/2\hbar}$$

be the standard Gaussian on $\mathbb{R}^n$. *We have*

$$\widehat{S}\phi_0(x) = e^{i\gamma}\left(\frac{1}{\pi\hbar}\right)^{n/4} (\det X)^{1/4} e^{-\frac{1}{2\hbar}(X+iY)x\cdot x} \tag{9.8}$$

where the phase γ is a real constant and X and Y are real symmetric matrices

$$X = (AA^T + BB^T)^{-1} \tag{9.9}$$

$$Y = -(CA^T + DB^T)(AA^T + BB^T)^{-1}. \tag{9.10}$$

Proof. We have $W(\widehat{S}\phi_0)(z) = W\phi_0(S^{-1}z)$. Since

$$W\phi_0(z) = \left(\frac{1}{\pi\hbar}\right)^n e^{-\frac{1}{\hbar}|z|},$$

we have

$$W(\widehat{S}\phi_0)(z) = \left(\frac{1}{\pi\hbar}\right)^n e^{-\frac{1}{\hbar}(S^{-1})^T S^{-1}z\cdot z}. \tag{9.11}$$

The inverse of S being given by

$$S = \begin{pmatrix} D^T & -B^T \\ -C^T & A^T \end{pmatrix},$$

an immediate calculation yields

$$(S^{-1})^T S^{-1} = \begin{pmatrix} CC^T + DD^T & -DB^T - CA^T \\ -BD^T - AC^T & AA^T + BB^T \end{pmatrix}.$$

Comparison with formulas (4.8) and (4.9) yields

$$X + YX^{-1}Y = CC^T + DD^T, \quad YX^{-1} = -DB^T - CA^T$$
$$X^{-1}Y = -BD^T - AC^T, \quad X^{-1} = AA^T + BB^T.$$

Solving this system of matrix equations yields the solutions (9.9) and (9.10). □

9.2.2 Pre-Iwasawa factorization of a symplectic matrix

Recall that every free symplectic matrix

$$S = \begin{pmatrix} A & B \\ C & D \end{pmatrix} \in \mathrm{Sp}(n), \quad \det B \neq 0$$

can be factorized as $S = V_{-P}M_L J V_{-Q}$ where

$$V_{-P} = \begin{pmatrix} I & 0 \\ P & I \end{pmatrix}, \quad M_L = \begin{pmatrix} L^{-1} & 0 \\ 0 & L^T \end{pmatrix}.$$

The pre-Iwasawa factorization is a variant of this result; it states that every $S \in \mathrm{Sp}(n)$ can be written as a product of an element of a certain subgroup $\mathrm{Sp}_0(n)$ of $\mathrm{Sp}(n)$ and of a symplectic rotation $U \in U(n)$.

Definition 106. We will denote by $\mathrm{Sp}_0(n)$ the subgroup of $\mathrm{Sp}(n)$ generated by the symplectic matrices V_{-P} and M_L. That subgroup $\mathrm{Sp}_0(n)$ is called the local symplectic group. The subgroup $\mathrm{Mp}_0(n) = (\pi^{\mathrm{Mp}})^{-1}(\mathrm{Sp}_0(n))$ of $\mathrm{Mp}(n)$ is called the local metaplectic group.

The group $\mathrm{Sp}_0(n)$ is the isotropy subgroup (= stabilizer) of the subspace $0 \oplus \mathbb{R}^n$ of $\mathbb{R}^{2n}$. It thus consists of all symplectic block matrices with upper corner $B = 0$, that is of matrices of the type

$$V_{-P}M_L = \begin{pmatrix} L^{-1} & 0 \\ PL^{-1} & L^T \end{pmatrix}, \quad M_L V_{-P} = \begin{pmatrix} L^{-1} & 0 \\ L^T P & L^T \end{pmatrix}.$$

The local metaplectic group $\mathrm{Mp}_0(n)$ consists of all products $\widehat{V}_{-P}\widehat{M}_{L,m}$ (or $\widehat{M}_{L,m}\widehat{V}_{-P}$); they are local operators (hence the terminology) in the sense that they do not increase the supports of the functions (or distributions) to which they are applied; intuitively, local operators do not contain any Fourier transform.

Proposition 107 (Iwasawa factorization). *For every $S \in \mathrm{Sp}(n)$, there exists a unique $S_0 \in \mathrm{Sp}_0(n)$ and a unique matrice $P = P^T$ and $L = L^T > 0$ and $U \in U(n)$ such that $S = S_0 U$; more precisely*

$$S_0 = \begin{pmatrix} I & 0 \\ P & I \end{pmatrix}\begin{pmatrix} L^{-1} & 0 \\ 0 & L \end{pmatrix}, \quad U = \begin{pmatrix} X & Y \\ -Y & X \end{pmatrix}$$

where P, L, X, Y are given by:

$$P = (CA^T + DB^T)(AA^T + BB^T)^{-1} = P^T \tag{9.12}$$

$$L = (AA^T + BB^T)^{-1/2} = L^T > 0 \tag{9.13}$$

$$X = (AA^T + BB^T)^{-1/2}A, \; Y = (AA^T + BB^T)^{-1/2}B. \tag{9.14}$$

Proof. It is purely computational writing

$$\begin{pmatrix} A & B \\ C & D \end{pmatrix} = \begin{pmatrix} I & 0 \\ P & I \end{pmatrix}\begin{pmatrix} L^{-1} & 0 \\ 0 & L \end{pmatrix}\begin{pmatrix} X & Y \\ -Y & X \end{pmatrix} \tag{9.15}$$

and determining the matrices P, L, X, Y in terms on A, B, C, D. □

9.2.3 Application to Gaussians

Let us return to formula (9.11)

$$W(\widehat{S}\phi_0)(z) = \left(\frac{1}{\pi\hbar}\right)^n e^{-\frac{1}{\hbar}(S^{-1})^T S^{-1} z \cdot z} \tag{9.16}$$

giving the Wigner transform of $\widehat{S} \in \mathrm{Mp}(n)$ applied to the standard Gaussian

$$\phi_0(x) = (\pi\hbar)^{-n/4} e^{-|x|^2/2\hbar}.$$

Let us focus on the product $(S^{-1})^T S^{-1} = (SS^T)^{-1}$. If we replace S with any other symplectic matrix $S'S'^T$ such that $SS^T = SS^T$, we can replace $W(\widehat{S}\phi_0)$ with $W(\widehat{S}'\phi_0)$, and the functions $\widehat{S}\phi_0$ and $\widehat{S}'\phi_0$ will thus be the same (up to an innocuous factor with modulus one). If we now use the Iwasawa factorization $S = S_0 U$, we have $SS^T = S_0 S_0^T$. Since

$$S_0 = \begin{pmatrix} I & 0 \\ P & I \end{pmatrix} \begin{pmatrix} L^{-1} & 0 \\ 0 & L \end{pmatrix} = \begin{pmatrix} L^{-1} & 0 \\ PL^{-1} & L \end{pmatrix},$$

we have

$$S_0 S_0^T = \begin{pmatrix} L^{-2} & L^{-2}P \\ PL^{-2} & PL^{-2}P + L^2 \end{pmatrix}.$$

More interesting is that we can directly obtain $\widehat{S}\phi_0$ (up to a sign) using this procedure: We have $\widehat{S}\phi_0 = \widehat{S}_0\phi_0$ where $\widehat{S}_0 = \widehat{V}_{-P}\widehat{M}_{L,m}$ that is, explicitly,

$$\widehat{S}_0\phi_0(x) = e^{\frac{i}{2\hbar}Px\cdot x} i^m \sqrt{|\det L|}\phi_0(Lx).$$

Since $L > 0$, we have $\det L > 0$ and $m \in \{0, 2\}$, hence

$$\widehat{S}_0\phi_0(x) = \pm e^{\frac{i}{2\hbar}Px\cdot x} \sqrt{\det L}\phi_0(Lx),$$

that is

$$\widehat{S}_0\phi_0(x) = \pm(\pi\hbar)^{-n/4} \sqrt{\det L} e^{-\frac{1}{2\hbar}(L^2+iP)x\cdot x}$$

with

$$P = (CA^T + DB^T)(AA^T + BB^T)^{-1} \tag{9.17}$$

$$L = (AA^T + BB^T)^{-1/2}. \tag{9.18}$$

9.3 Symplectic covariance of Weyl operators

Weyl operators are the *only* pseudo-differential operator calculus for which symplectic covariance holds. See the Comments and References at the end of the chapter.

We have just proven that $\widehat{D}(z_0)$ and $\widehat{R}(z_0)$ satisfy the symplectic covariance formulas

$$\widehat{S}\widehat{D}(z)\widehat{S}^{-1} = \widehat{D}(Sz), \quad \widehat{S}\widehat{R}(z)\widehat{S}^{-1} = \widehat{R}(Sz)$$

for $\widehat{S} \in \mathrm{Mp}(n)$ and $S = \pi^{\mathrm{Mp}}(\widehat{S})$ for every $z \in \mathbb{R}^{2n}$. Also, for all $(\psi, \phi) \in L^2(\mathbb{R}^n) \times L^2(\mathbb{R}^n)$, the cross-Wigner distribution satisfies

$$W(\widehat{S}\psi, \widehat{S}\phi)(z) = W(\psi, \phi)(S^{-1}z) \tag{9.19}$$

and, similarly, for the cross-ambiguity function,

$$A(\widehat{S}\psi, \widehat{S}\phi)(z) = A(\psi, \phi)(S^{-1}z). \tag{9.20}$$

Proposition 108. *Let $\widehat{S} \in \mathrm{Mp}(n)$ have projection $S \in \mathrm{Sp}(n)$ and be any of the two metaplectic operators with $S = \pi^{\mathrm{Mp}}(\widehat{S})$. For every Weyl operator $\mathrm{Op_W}(a)$, we have*

$$\mathrm{Op_W}(a \circ S) = \widehat{S}^{-1}\, \mathrm{Op_W}(a)\widehat{S}. \tag{9.21}$$

Proof. Let us denote $\widehat{B}$ the Weyl operator with symbol $a \circ S$. We have

$$\widehat{B}\psi = \int_{\mathbb{R}^{2n}} a_\sigma(Sz)\widehat{D}(z)\psi dz$$

that is, performing the change of variables $Sz \longmapsto z$ and taking into account the fact that $\det S = 1$,

$$\widehat{B}\psi = \int_{\mathbb{R}^{2n}} a_\sigma(z)\widehat{D}(S^{-1}z)\psi dz.$$

Since $\widehat{S}^{-1}\widehat{D}(z)\widehat{S} = \widehat{D}(S^{-1}z)$, we have

$$\begin{aligned}\widehat{B}\psi &= \int_{\mathbb{R}^{2n}} a_\sigma(z)\widehat{S}^{-1}\widehat{D}(z)\widehat{S}\psi dz \\ &= \widehat{S}^{-1}\Big(\int_{\mathbb{R}^{2n}} a_\sigma(z)\widehat{D}(z)dz\Big)\widehat{S}\psi\end{aligned}$$

which is (9.21). □

Notice that this result could be proven as well using the alternative (9.21) definition

$$\mathrm{Op_W}(a) = \left(\frac{1}{\pi\hbar}\right)^n \int_{\mathbb{R}^{2n}} a(z_0)\widehat{R}(z_0)dz_0,$$

expressing the Weyl correspondence in terms of the Grossmann–Royer operators $\psi = \phi$.

9.4 Maximality of the symplectic group

We are going to show that the symplectic group is in a sense “almost” maximal for covariance properties. More precisely, we will show that if $M \in GL(2n,\mathbb{R})$, then, for every $\psi \in \mathcal{S}'(\mathbb{R}^n)$ there exists $\psi' \in \mathcal{S}'(\mathbb{R}^n)$ such that

$$W\psi \circ M = W\psi'$$

if and only if M is either symplectic or antisymplectic (i. e., $M^TJM = -J$).

9.4.1 A technical lemma

Recall that an automorphism M of $\mathbb{R}^{2n}$ is antisymplectic if $\sigma(Mz, Mz') = -\sigma(z,z')$ for all $z,z' \in \mathbb{R}^{2n}$; in matrix notation, $M^TJM = -J$. Equivalently, $CM \in \operatorname{Sp}(n)$ where $C = \begin{pmatrix} I & 0 \\ 0 & -I \end{pmatrix}$.

Lemma 109. *Let $M \in GL(2n,\mathbb{R})$, and assume that $M^TGM \in \operatorname{Sp}(n)$ for every*

$$G = \begin{pmatrix} X & 0 \\ 0 & X^{-1} \end{pmatrix} \in \operatorname{Sp}^+(n). \tag{9.22}$$

Then M is either symplectic, or anti-symplectic.

Proof. We first remark that, taking $G = I$ in the condition $M^TGM \in \operatorname{Sp}(n)$, we have $M^TM \in \operatorname{Sp}(n)$. Next, we can write $M = HP$ where $H = M(M^TM)^{-1/2}$ is orthogonal and $P = (M^TM)^{1/2} \in \operatorname{Sp}^+(n)$ (polar decomposition theorem). It follows that the condition $M^TGM \in \operatorname{Sp}(n)$ is equivalent to $P(H^TGH)P \in \operatorname{Sp}(n)$; since P is symplectic, so is P^{-1}, and hence $H^TGH \in Sp(n)$ for all G of the form (9.22). Let us now make the following particular choice for G by taking

$$G = \begin{pmatrix} \Lambda & 0 \\ 0 & \Lambda^{-1} \end{pmatrix}, \quad \Lambda = \operatorname{diag}(\lambda_1,\dots,\lambda_n)$$

with $\lambda_j > 0$ for $1 \leq j \leq n$. We thus have

$$H^T \begin{pmatrix} \Lambda & 0 \\ 0 & \Lambda^{-1} \end{pmatrix} H \in \operatorname{Sp}^+(n)$$

for every Λ of this form. Let $U \in U(n)$ be a symplectic rotation such that

$$H^T \begin{pmatrix} \Lambda & 0 \\ 0 & \Lambda^{-1} \end{pmatrix} H = U^T \begin{pmatrix} \Lambda & 0 \\ 0 & \Lambda^{-1} \end{pmatrix} U\,;$$

we have $H^TGH \in \operatorname{Sp}^+(n)$ and the eigenvalues of H^TGH are those of G since H is orthogonal. Setting $R = HU^T$, the previous equality is equivalent to

$$\begin{pmatrix} \Lambda & 0 \\ 0 & \Lambda^{-1} \end{pmatrix} R = R \begin{pmatrix} \Lambda & 0 \\ 0 & \Lambda^{-1} \end{pmatrix}. \tag{9.23}$$

Writing $R = \left(\begin{smallmatrix} A & B \\ C & D \end{smallmatrix}\right)$, we get the conditions

$$\Lambda A = A\Lambda, \quad \Lambda B = B\Lambda^{-1}$$
$$\Lambda^{-1}C = C\Lambda, \quad \Lambda^{-1}D = D\Lambda^{-1}$$

for all Λ. It follows that A and D must themselves be diagonal $A = \operatorname{diag}(a_1, \dots, a_n)$, $D = \operatorname{diag}(d_1, \dots, d_n)$. On the other hand, choosing $\Lambda = \lambda I$, $\lambda \neq 1$, we get $B = C = 0$. Hence, taking into account the fact that $R \in O(2n, \mathbb{R})$, we must have

$$R = \begin{pmatrix} A & 0 \\ 0 & D \end{pmatrix}, \quad A^2 = D^2 = I. \tag{9.24}$$

Conversely, if R is of the form (9.24), then (9.23) holds for any positive-definite diagonal Λ. We conclude that M has to be of the form $M = RUP$ where R is of the form (9.24). Since $UP \in \operatorname{Sp}(n)$, the relation ship $M^T GM \in \operatorname{Sp}(n)$ implies $RGR \in \operatorname{Sp}(n)$. For each pair i, j with $1 \leq i < j \leq n$, we now choose the following matrix X in (9.22):

$$X^{(ij)} = I + \frac{1}{2}E^{(ij)} \tag{9.25}$$

where $E^{(ij)}$ is the symmetric matrix whose entries are all zero except the ones on the i-th row and j-th column and on the j-th row and i-th column which are equal to one. A simple calculation then shows that

$$RGR = \begin{pmatrix} AX^{(ij)}A & 0 \\ 0 & D(X^{(ij)})^{-1}D \end{pmatrix}, \tag{9.26}$$

and if we impose the condition $RGR \in \operatorname{Sp}(n)$, we obtain

$$AX^{(ij)}AD(X^{(ij)})^{-1}D = I \Leftrightarrow X^{(ij)}AD = ADX^{(ij)}. \tag{9.27}$$

In other words, the matrix AD commutes with every real positive-definite $n \times n$ matrix $X^{(ij)}$ of the form (9.25). Let us write $AD = \operatorname{diag}(c_1, \dots, c_n)$ with $c_j = a_j d_j$ for $1 \leq j \leq n$. Then if we apply (9.27) to (9.25) for $i < j$, we conclude that $c_i = c_j$ which means that the entries of the matrix AD are all equal, that is, either $AD = I$ or $AD = -I$, or equivalently $A = D$ or $A = -D$. In the first case, R is symplectic and so is M. In the second case, R is anti-symplectic; but then M is also anti-symplectic. □

9.4.2 The maximality property for the Wigner transform

The considerations just posed allow us to prove the main result of this section:

Proposition 110. *Let $M \in GL(2n, \mathbb{R})$.* (i) *Assume that M is antisymplectic: $S = CM \in$* Sp(*n*) *where $C = \begin{pmatrix} I & 0 \\ 0 & -I \end{pmatrix}$; then for every $\psi \in \mathcal{S}'(\mathbb{R}^n)$,*

$$W\psi(Mz) = W(\widehat{S}^{-1}\overline{\psi})(z) \tag{9.28}$$

where $\widehat{S}$ is any of the two elements of the metaplectic group Mp(*n*) *covering S.* (ii) *Conversely, assume that for any $\psi \in \mathcal{S}(\mathbb{R}^n)$ there exists $\psi' \in \mathcal{S}'(\mathbb{R}^n)$ such that*

$$W\psi(Mz) = W\psi'(z). \tag{9.29}$$

Then M is either symplectic or antisymplectic.

Proof. (i) It is sufficient to assume that $\psi \in \mathcal{S}(\mathbb{R}^n)$. We have

$$\begin{aligned} W\psi(Cz) &= \left(\frac{1}{2\pi\hbar}\right)^n \int_{\mathbb{R}^n} e^{\frac{i}{\hbar}p\cdot y}\psi\left(x + \frac{1}{2}y\right)\overline{\psi}\left(x - \frac{1}{2}y\right)dy \\ &= \left(\frac{1}{2\pi\hbar}\right)^n \int_{\mathbb{R}^n} e^{-\frac{i}{\hbar}p\cdot y}\psi\left(x - \frac{1}{2}y\right)\overline{\psi}\left(x + \frac{1}{2}y\right)dy \\ &= W\overline{\psi}(z). \end{aligned}$$

It follows that

$$W\psi(Mz) = W\psi(CSz) = W\overline{\psi}(Sz),$$

hence formula (9.28). (ii) Choosing for ψ a Gaussian

$$\psi_X(x) = \left(\frac{1}{\pi\hbar}\right)^{n/4} (\det X)^{1/4} e^{-\frac{1}{2\hbar}Xx\cdot x} \tag{9.30}$$

(X real symmetric and positive definite) we have

$$W\psi_X(z) = \left(\frac{1}{\pi\hbar}\right)^n e^{-\frac{1}{\hbar}Gz\cdot z} \tag{9.31}$$

where

$$G = \begin{pmatrix} X & 0 \\ 0 & X^{-1} \end{pmatrix} \tag{9.32}$$

is positive definite and belongs to Sp(n). Condition (9.29) implies that we must have

$$W\psi'(z) = \left(\frac{1}{\pi\hbar}\right)^n e^{-\frac{1}{\hbar}M^T GMz\cdot z}.$$

The Wigner transform of a function is a Gaussian if and only if the function itself is a Gaussian (see [30, 31]). This can be seen in the following way: The matrix M^TGM being symmetric and positive definite, we can use a Williamson diagonalization: There exists $S \in \operatorname{Sp}(n)$ such that

$$S^T(M^TGM)S = \Delta = \begin{pmatrix} \Sigma & 0 \\ 0 & \Sigma \end{pmatrix}, \tag{9.33}$$

and hence

$$W\psi'(Sz) = \left(\frac{1}{\pi\hbar}\right)^n e^{-\frac{1}{\hbar}\Delta z\cdot z}.$$

In view of the symplectic covariance of the Wigner transform, we have

$$W\psi'(Sz) = W\psi''(z), \quad \psi'' = \widehat{S}^{-1}\psi'$$

where $\widehat{S} \in \operatorname{Mp}(n)$ is one of the two elements of the metaplectic group covering S. We now show that the equality

$$W\psi''(z) = \left(\frac{1}{\pi\hbar}\right)^n e^{-\frac{1}{\hbar}\Delta z\cdot z} \tag{9.34}$$

implies that ψ'' must be a Gaussian of the form (4.2), and hence $W\psi''$ must be of the type (9.31, 9.32). That ψ'' must be a Gaussian follows from $W\psi'' \geq 0$ and Hudson's theorem (see e. g., [28]). If ψ'' were of the more general type

$$\psi_{X,Y}(x) = \left(\frac{1}{\pi\hbar}\right)^{n/4} (\det X)^{1/4} e^{-\frac{1}{2\hbar}(X+iY)x\cdot x} \tag{9.35}$$

(X, Y are real and symmetric and X is positive definite), the matrix G in (9.31) would be

$$G = \begin{pmatrix} X + YX^{-1}Y & YX^{-1} \\ X^{-1}Y & X^{-1} \end{pmatrix} \tag{9.36}$$

which is only compatible with (9.34) if $Y = 0$. In addition, due to the parity of $W\psi''$, ψ'' must be even hence Gaussians more general than $\psi_{X,Y}$ are excluded. It follows from these considerations that we have

$$\Delta = \begin{pmatrix} \Sigma & 0 \\ 0 & \Sigma \end{pmatrix} = \begin{pmatrix} X & 0 \\ 0 & X^{-1} \end{pmatrix},$$

so that $\Sigma = \Sigma^{-1}$. Since $\Sigma > 0$, this implies that we must have $\Sigma = I$, and hence, using formula (9.33), $S^T(M^TGM)S = I$. It follows that we must have $M^TGM \in \operatorname{Sp}(n)$ for every $G = \begin{pmatrix} X & 0 \\ 0 & X^{-1} \end{pmatrix} \in \operatorname{Sp}^+(n)$. In view of Lemma 109, the matrix M must then be either symplectic or antisymplectic. □

9.4.3 Maximal covariance of Weyl operators

The previous result implies the following maximal covariance result for Weyl operators:

Corollary 111. *Let $M \in GL(2n, \mathbb{R})$. Assume that there exists a unitary operator $\widehat{M} : L^2(\mathbb{R}^n) \longrightarrow L^2(\mathbb{R}^n)$ such that*

$$\widehat{M}\,\mathrm{Op_W}(a)\widehat{M}^{-1} = \mathrm{Op_W}(a \circ M^{-1}) \tag{9.37}$$

for all $a \in \mathcal{S}(\mathbb{R}^{2n})$. Then M is either symplectic or antisymplectic.

Proof. Suppose that (9.37) holds; then

$$\begin{aligned}(\widehat{M}\,\mathrm{Op_W}(a)\widehat{M}^{-1}\psi|\phi)_{L^2} &= \langle a \circ M^{-1}, W(\psi,\phi)\rangle \\ &= \langle a, W(\psi,\phi) \circ M\rangle.\end{aligned}$$

On the other hand, using the unitarity of $\widehat{M}$,

$$\begin{aligned}(\widehat{M}\,\mathrm{Op_W}(a)\widehat{M}^{-1}\psi|\phi)_{L^2} &= (\mathrm{Op_W}(a)\widehat{M}^{-1}\psi|\widehat{M}^{-1}\phi)_{L^2} \\ &= \langle a, W(\widehat{M}^{-1}\psi, \widehat{M}^{-1}\phi)\rangle.\end{aligned}$$

It follows that we must have

$$\langle\langle a, W(\psi,\phi) \circ M\rangle\rangle = \langle\langle a, W(\widehat{M}^{-1}\psi, \widehat{M}^{-1}\phi)\rangle\rangle$$

for all $(\psi,\phi) \in \mathcal{S}(\mathbb{R}^n) \times \mathcal{S}(\mathbb{R}^n)$, and hence, in particular, taking $\psi = \phi$:

$$\langle a, W\psi \circ M\rangle = \langle a, W(\widehat{M}^{-1}\psi)\rangle$$

for all $\psi \in \mathcal{S}(\mathbb{R}^n)$. Since a is arbitrary, this implies that we must have $W\psi \circ M = W(\widehat{M}^{-1}\psi)$. In view of Proposition 110, the automorphism M must be either symplectic or antisymplectic. □

Summarizing:

Proposition 112. *Let $a \longmapsto \mathrm{Op_W}(a)$ be a continuous linear mapping from $\mathcal{S}'(\mathbb{R}^{2n})$ on the space $\mathcal{L}(\mathcal{S}(\mathbb{R}^n), \mathcal{S}'(\mathbb{R}^n))$. Assume that:* (i) *If a only depends on $x \in \mathbb{R}^n$ and $a \in L^\infty(\mathbb{R}^n)$, then $\mathrm{Op_W}(a)$ is multiplication by $a(x)$;* (ii) *if $S \in \mathrm{Sp}(n)$, then $\mathrm{Op_W}(a \circ S) = \widehat{S}\,\mathrm{Op_W}(a)\widehat{S}^{-1}$. Then $a \longmapsto \mathrm{Op_W}(a)$ is the Weyl correspondence: $\mathrm{Op_W}(a) = \mathrm{Op_W}(a)$.*

9.4.4 Born–Jordan operators

The Born Jordan correspondence $a \longrightarrow \mathrm{Op_{BJ}}(a)$ is not identical to the Weyl correspondence $a \longrightarrow \mathrm{Op_W}(a)$, hence we cannot expect it to enjoy full symplectic covariance since the latter is a characteristic property of Weyl quantization. We have however the following partial result. Recall that the metaplectic group Mp(n) is generated by the modified Fourier transform $\widehat{J} = i^{-n/2}F$, the multiplication operators

$$\widehat{V}_{-P}\psi = e^{iPx^2/2\hbar}\psi \quad (P = P^T),$$

and the unitary scaling operators

$$\widehat{M}_{L,m}\psi(x) = i^m\sqrt{|\det L|}\psi(Lx)$$

($\det L \neq 0$, $m\pi = \arg\det L$). The projections of these operators on Sp(n) are, respectively, the symplectic matrix J and

$$V_{-P} = \begin{pmatrix} I & 0 \\ P & I \end{pmatrix}, \quad M_L = \begin{pmatrix} L^{-1} & 0 \\ 0 & L^2 \end{pmatrix}.$$

Proposition 113. *Let $a \in \mathcal{S}'(\mathbb{R}^{2n})$. We have*

$$\widehat{S}\,\mathrm{Op_{BJ}}(a)\widehat{S}^{-1} = \mathrm{Op_{BJ}}(a \circ S^{-1}) \tag{9.38}$$

for every $\widehat{S} \in$ Mp(n) which is a product of a (finite number) of operators $\widehat{J}$ and $\widehat{M}_{L,m}$.

Proof. It suffices to prove formula (9.38) for $\widehat{S} = \widehat{J}$ and $\widehat{S} = \widehat{M}_{L,m}$. Let first $\widehat{S}$ be an arbitrary element of Mp(n); we have

$$\begin{aligned} \widehat{S}\,\mathrm{Op_{BJ}}(a) &= \left(\frac{1}{2\pi\hbar}\right)^n \int a_\sigma(z)\Theta(z)\widehat{S}\widehat{T}(z)d^{2n}z \\ &= \left[\left(\frac{1}{2\pi\hbar}\right)^n \int a_\sigma(z)\Theta(z)\widehat{T}(Sz)d^{2n}z\right]\widehat{S} \end{aligned}$$

where the second equality follows from the usual symplectic covariance property $\widehat{S}\widehat{T}(z) = \widehat{T}(Sz)\widehat{S}$ of the Heisenberg operators. Making the change of variables $z' = Sz$ in the integral, we get, since $\det S = 1$,

$$\int a_\sigma(z)\Theta(z)\widehat{T}(Sz)d^{2n}z = \int a_\sigma(S^{-1}z)\Theta(S^{-1}z)\widehat{T}(z)d^{2n}z.$$

Now, by definition of the symplectic Fourier transform, we have

$$\begin{aligned} a_\sigma(S^{-1}z) &= \left(\frac{1}{2\pi\hbar}\right)^n \int e^{-\frac{i}{\hbar}\sigma(S^{-1}z,z')}a(z')d^{2n}z' \\ &= (a \circ S^{-1})_\sigma(z). \end{aligned}$$

Choosing $\widehat{S} = \widehat{M}_{L,m}$ we have

$$\Theta(M_L^{-1}z) = \frac{\sin(2\pi Lp \cdot (L^T)^{-1}x)}{2\pi Lp \cdot (L^T)^{-1}x} = \Theta(z);$$

similarly $\Theta(J^{-1}z) = \Theta(z)$, hence in both cases

$$\widehat{S}\,\mathrm{Op_{BJ}}(a) = \left(\int (a \circ S^{-1})_\sigma \Theta(z)\widehat{T}(z)d^{2n}z\right)\widehat{S}$$
$$= \mathrm{Op_{BJ}}(a \circ S^{-1})\widehat{S}$$

whence formula (9.38). □

This proof shows that the essential step consists in noting that $\Theta(S^{-1}z) = \Theta(z)$ when $S = J$ or $S = M_L$. It is clear that this property fails if one takes any operator $S = V_P$ with $P \neq 0$, so we cannot expect to have full symplectic covariance for Born–Jordan operators since the symplectic group $\mathrm{Sp}(n)$ is generated by the set of all matrices J, M_L and V_P.

9.5 Comments and references

The proof of the maximality results in Section 9.4 was first published in a joint work [19] with Nuno Dias and Joao Prata (the proof of Lemma 109 is due to Dias and Prata). That the Weyl transform is the only quantization that is a fully symplectically covariant was proven by Stein in his treatise [68]; Stein's argument has been detailed by Wong [76].

10 The Feichtinger algebra

The Feichtinger algebra $S_0(\mathbb{R}^n)$ is a normed Banach algebra of functions that van be defined in terms of the Wigner transform. It is a good substitute for the Schwartz space $\mathcal{S}(\mathbb{R}^n)$ as long as one is not interested in differentiation properties. It is moreover the smallest Banach space containing $\mathcal{S}(\mathbb{R}^n)$ and being invariant under the action of the metaplectic group and the Heisenberg displacement. The dual space $S_0'(\mathbb{R}^n)$ of $S_0(\mathbb{R}^n)$ contains many basic distributions such as the Dirac distribution δ or its translates. These properties, together with the fact that Banach spaces are mathematically easier to deal with than Fréchet spaces, makes the Feichtinger algebra into a tool of choice in harmonic analyses and its applications to quantum mechanics.

10.1 Definition and first properties

10.1.1 Definition of the Feichtinger algebra

In what follows, ϕ will be a non-zero element of $\mathcal{S}(\mathbb{R}^n)$ with L^2 norm equal to one, and referred to as a "window".

Definition 114. The Feichtinger algebra $S_0(\mathbb{R}^n)$ consists of all functions $\psi \in L^2(\mathbb{R}^n)$ such that $W(\psi, \phi) \in L^1(\mathbb{R}^{2n})$ for every window ϕ. The number

$$\|\psi\|_{\phi,S_0} = \|W(\psi,\phi)\|_{L^1(\mathbb{R}^{2n})} = \int_{\mathbb{R}^{2n}} |W(\psi,\phi)(z)|dz \tag{10.1}$$

is called the norm of ψ relative to the window ϕ.

Admittedly, this definition is not very practical, because it means that we should in principle verify an uncountable set of conditions. We will see later that it actually suffices to verify that $W(\psi, \phi) \in L^1(\mathbb{R}^{2n})$ for only *one* window ϕ.

That $S_0(\mathbb{R}^n)$ is a vector space is clear. That $\|\cdot\|_{\phi,S_0}$ indeed is a norm on $S_0(\mathbb{R}^n)$ for each $\phi \in \mathcal{S}(\mathbb{R}^n)$, $\phi \neq 0$ is also obvious. For instance, $\|\psi\|_{\phi,S_0} = 0$ is equivalent to $W(\psi, \phi) = 0$, that is to $\psi = 0$ since $\phi \neq 0$ the triangle inequality follows from the sesquilinearity of W and so does the property $\|\lambda\psi\|_{\phi,S_0} = |\lambda| \, \|\psi\|_{\phi,S_0}$.

Here is an elementary example of an element of an element of $S_0(\mathbb{R}^n)$: the continuous but not differentiable function

$$\psi(x) = \begin{cases} 1 - |x| & \text{if } |x| \leq 1 \\ 0 & \text{if } |x| > 1. \end{cases}$$

That the definition of the Feichtinger algebra is independent of the choice of window ϕ – as the definition suggests – is not quite obvious. Let us prove this property in detail.

https://doi.org/10.1515/9783110722772-010

Proposition 115. (i) *Let $\psi \in L^2(\mathbb{R}^n)$. We have $\psi \in S_0(\mathbb{R}^n)$ if and only if there exists one window ϕ such that $W(\psi,\phi) \in L^1(\mathbb{R}^{2n})$.* (ii) *The norms $\|\cdot\|_{\phi,S_0}$ are all equivalent when ϕ ranges over $\mathcal{S}(\mathbb{R}^n)$ and thus define the same topology on $S_0(\mathbb{R}^n)$.*

Proof. Let us express ψ in terms of $W(\psi,\phi)$ using the reconstruction formula (3.17):

$$\psi(x) = \frac{2^n}{(\gamma|\phi)_{L^2}} \int_{\mathbb{R}^{2n}} W(\psi,\phi)(z_0)\widehat{R}(z_0)\gamma(x)dz_0 \tag{10.2}$$

where $\gamma \in \mathcal{S}(\mathbb{R}^n)$ is such that $(\gamma|\phi)_{L^2} \neq 0$. Applying $\widehat{R}(z')$ to both sides of this equality, we get, in view of the product formula (2.11) for reflection operators,

$$\widehat{R}(z')\psi = \frac{2^n}{(\gamma|\phi)_{L^2}} \int_{\mathbb{R}^{2n}} W(\psi,\phi)(z)e^{\frac{2i}{\hbar}\sigma(z,z')}\widehat{D}(2z-2z')\gamma dz.$$

Let ϕ' be a second window; by definition of the cross-Wigner transform, we have

$$(\widehat{R}(z')\psi|\phi')_{L^2} = (\pi\hbar)^n W(\psi,\phi')(z')$$

and by definition (3.18) of the cross-ambiguity function

$$\begin{aligned}(\widehat{D}(2z-2z')\gamma|\phi')_{L^2} &= (2\pi\hbar)^n A(\gamma,\phi')(2z'-2z)\\ &= (\pi\hbar)^n W(\gamma,\phi'^{\vee})(z'-z)\end{aligned}$$

where $\phi'^{\vee}(x) = \phi'(-x)$, the second equality following from formula (3.22) relating cross-ambiguity and cross-Wigner transforms. Formula (10.2) thus yields

$$W(\psi,\phi')(z') = \frac{2^n}{(\gamma|\phi)_{L^2}} \int_{\mathbb{R}^{2n}} W(\psi,\phi)(z)e^{\frac{2i}{\hbar}\sigma(z,z')n}W(\gamma,\phi'^{\vee})(z'-z)dz,$$

and hence

$$|W(\psi,\phi')(z')| \leq \frac{2^n}{|(\gamma|\phi)_{L^2}|} \int_{\mathbb{R}^{2n}} |W(\psi,\phi)(z)||W(\gamma,\phi'^{\vee})(z'-z)|dz,$$

that is

$$|W(\psi,\phi')| \leq \frac{2^n}{|(\gamma|\phi)_{L^2}|}|W(\psi,\phi)| * |W(\gamma,\phi'^{\vee})|. \tag{10.3}$$

Integrating both sides of this inequality with respect to z yields

$$\|\psi\|_{\phi',S_0} \leq \frac{2^n}{(\gamma|\phi)_{L^2}}\|\psi\|_{\phi,S_0}\|\gamma\|_{\phi'^{\vee},S_0} \tag{10.4}$$

where we have used the inequality $\|F * G\|_{L^1} \leq \|F\|_{L^1}\|G\|_{L^1}$ valid for any integrable functions F and G. The equivalence of the norms $\|\cdot\|_{\psi,S_0}$ and $\|\cdot\|_{\phi,S_0}$ follows. □

10.1.2 First properties

We are going to show that it turns out that the window ϕ used in the definition of $S_0(\mathbb{R}^n)$ can itself be chosen in $S_0(\mathbb{R}^n)$. As a consequence, we can redefine the Feichtinger algebra in terms of the Wigner transform of its elements: A square integrable function ψ is in $S_0(\mathbb{R}^n)$ if and only if $W\psi \in L^1(\mathbb{R}^{2n})$. To prove this important result, we will need the following technical lemma:

Lemma 116. *Let $\phi \in \mathcal{S}(\mathbb{R}^n)$, $\phi \neq 0$. The following properties are equivalent:* (i) $\psi \in \mathcal{S}(\mathbb{R}^n)$; (ii) $W(\psi, \phi) \in \mathcal{S}(\mathbb{R}^{2n})$; (iii) *for every $N \geq 0$, there exists $C_N \geq 0$ such that*

$$|W(\psi, \phi)| \leq C_N (1 + |z|)^{-N}.$$

Proof. It is clear, by definition of $\mathcal{S}(\mathbb{R}^n)$ and $\mathcal{S}(\mathbb{R}^{2n})$, that (i)$\Longrightarrow$(ii)$\Longrightarrow$(iii). Let us prove that (iii)$\Longrightarrow$(i). It is easily verified that the function χ defined by

$$\chi(x) = \frac{2^n}{\|\phi\|_{L^2}} \int_{\mathbb{R}^{2n}} W(\psi, \phi)(z) \widehat{R}(z) \phi(x) dz$$

is in $\mathcal{S}(\mathbb{R}^n)$; but then $\chi = \psi$ in view of the proof of Proposition 40 since (iii) implies in particular that $W(\psi, \phi) \in L^2(\mathbb{R}^{2n})$. Hence $\psi \in \mathcal{S}(\mathbb{R}^n)$ as we set out to prove. □

Proposition 117. *Let $(\psi, \phi) \in L^2(\mathbb{R}^n) \times L^2(\mathbb{R}^n)$.* (i) *If $W(\psi, \phi) \in L^1(\mathbb{R}^{2n})$, then both ψ and ϕ are in $S_0(\mathbb{R}^n)$;* (ii) *we have $\psi \in S_0(\mathbb{R}^n)$ if and only if $W(\psi, \phi) \in L^1(\mathbb{R}^{2n})$ for one (and hence every) $\phi \in S_0(\mathbb{R}^n)$;* (iii) *a function $\psi \in L^2(\mathbb{R}^n)$ belongs to $S_0(\mathbb{R}^n)$ if and only if $W\psi \in L^1(\mathbb{R}^{2n})$.*

Proof. Property (ii) immediately follows from (i). Let us prove property (i). The condition that $\psi, \phi \in L^2(\mathbb{R}^n)$ implies that $W(\psi, \phi)$ is a square-integrable and continuous function. Recall that, in the course of the proof of Proposition 115, we proved the inequality (10.3).

$$|W(\psi, \phi')| \leq \frac{2^n}{|(\gamma|\phi)_{L^2}|} |W(\psi, \phi)| * |W(\gamma, \phi'^\vee)|.$$

Choosing $\gamma = \phi'^\vee$, this inequality becomes

$$|W(\psi, \phi')| \leq \frac{2^n}{|(\gamma|\phi)_{L^2}|} |W(\psi, \phi)| * |W(\phi'^\vee)|,$$

hence, integrating both sides,

$$\|\psi\|_{\phi', S_0} \leq \frac{2^n}{|(\gamma|\phi)_{L^2}|} \|W(\psi, \phi)\|_{L^1} \|W\phi'^\vee\|_\infty < \infty$$

which shows that $\psi \in S_0(\mathbb{R}^n)$. Swapping ψ and ϕ, this inequality becomes

$$\|\phi\|_{\phi', S_0} \leq \frac{2^n}{|(\gamma|\phi)_{L^2}|} \|W(\phi, \psi)\|_{L^1} \, \|W(\phi'^\vee)\|_\infty < \infty,$$

hence we also have $\phi \in S_0(\mathbb{R}^n)$. (iii) In view of (i), the condition $W\psi \in L^1(\mathbb{R}^{2n})$ implies that $\psi \in S_0(\mathbb{R}^n)$. If conversely $\psi \in S_0(\mathbb{R}^n)$, then $W\psi \in L^1(\mathbb{R}^{2n})$ in view of (ii). □

The Feichtinger algebra $S_0(\mathbb{R}^n)$ is a subspace of several usual spaces of functions:

Proposition 118. *We have the chain of inclusions*

$$S_0(\mathbb{R}^n) \subset C^0(\mathbb{R}^n) \cap L^1(\mathbb{R}^n) \cap F(L^1(\mathbb{R}^n)). \tag{10.5}$$

Proof. We will use again the reconstruction formula

$$\psi(x) = \frac{2^n}{(\gamma|\phi)_{L^2}} \int_{\mathbb{R}^{2n}} W(\psi,\phi)(z_0)\widehat{R}(z_0)\gamma(x)dz_0$$

which is valid for all $\gamma \in \mathcal{S}(\mathbb{R}^n)$ such that $(\gamma|\phi)_{L^2} \neq 0$. Putting $\Delta\psi(x) = \psi(x+\Delta x) - \psi(x)$, we have

$$\begin{aligned}
|\Delta\psi(x)| &\leq \frac{2^n}{|(\gamma|\phi)_{L^2}|} \int_{\mathbb{R}^{2n}} |W(\psi,\phi)(z_0)||\widehat{R}(z_0)(\gamma(x+\Delta x) - \gamma(x))|dz_0 \\
&\leq \frac{2^n}{|(\gamma|\phi)_{L^2}|} \|W(\psi,\phi)\|_{L^1} \sup_{z_0} |\widehat{R}(z_0)(\gamma(x+\Delta x) - \gamma(x))| \\
&= \frac{2^n}{|(\gamma|\phi)_{L^2}|} \|W(\psi,\phi)\|_{L^1} \sup_{x_0} |(\gamma(2x_0 - x - \Delta x) - \gamma(2x_0 - x))|
\end{aligned}$$

where the last equality follows from the definition of $\widehat{R}(z)$, from which readily follows that $\lim_{\Delta x\to 0} \Delta\psi = 0$; hence, $S_0(\mathbb{R}^n) \subset C^0(\mathbb{R}^n)$. Let us next show that $S_0(\mathbb{R}^n) \subset L^1(\mathbb{R}^n)$. Let $\psi \in S_0(\mathbb{R}^n)$. Using again the reconstruction formula, we get

$$\begin{aligned}
|\psi(x)| &\leq \frac{2^n}{|(\gamma|\phi)_{L^2}|} \int_{\mathbb{R}^{2n}} |W(\psi,\phi)(z_0)||\gamma(2x_0 - x)|dz_0 \\
&\leq \frac{2^n}{|(\gamma|\phi)_{L^2}|} \int_{\mathbb{R}^{2n}} |W(\psi,\phi)(z_0)||\gamma(2x_0 - x)|dz_0,
\end{aligned}$$

and hence, integrating in x,

$$\|\psi\|_{L^1} \leq \frac{2^n}{(\gamma|\phi)_{L^2}} \|\psi\|_{\phi,S_0} \|\gamma\|_{L^\infty} < \infty$$

so that $\psi \in S_0(\mathbb{R}^n)$. To prove the inclusion $S_0(\mathbb{R}^n) \subset F(L^1(\mathbb{R}^n))$, it suffices to note that $S_0(\mathbb{R}^n)$ is invariant under a Fourier transform: we have $\psi \in S_0(\mathbb{R}^n)$ if and only if $W(\psi, F^{-1}\phi) \in L^1(\mathbb{R}^{2n})$ for every window ϕ since F is an automorphism $\mathcal{S}(\mathbb{R}^n) \longrightarrow \mathcal{S}(\mathbb{R}^n)$. Now,

$$W(F^{-1}\psi, F^{-1}\phi)(z) = W(\psi,\phi)(Jz) \tag{10.6}$$

in view of the symplectic covariance property of the cross-Wigner transform. It follows from the inclusion $S_0(\mathbb{R}^n) \subset L^1(\mathbb{R}^n)$ that we have $F^{-1}\psi \in L^1(\mathbb{R}^n)$, and hence $\psi \in F(L^1(\mathbb{R}^n))$ as claimed. □

The space $\mathcal{S}(\mathbb{R}^n)$ is dense in $S_0(\mathbb{R}^n)$:

Proposition 119. *We have $\mathcal{S}(\mathbb{R}^n) \subset S_0(\mathbb{R}^n)$, and $\mathcal{S}(\mathbb{R}^n)$ is dense in $S_0(\mathbb{R}^n)$.*

Proof. Let us show that $\mathcal{S}(\mathbb{R}^n) \subset S_0(\mathbb{R}^n)$. Let $\psi \in \mathcal{S}(\mathbb{R}^n)$; for every window ϕ, we have $W(\psi,\phi) \in \mathcal{S}(\mathbb{R}^{2n})$, hence for every $N > 0$ there exists $C_N > 0$ such that

$$|W(\psi,\phi)(z)| \leq C_N(1+|z|)^{-N}.$$

It follows, by definition of the norm $\|\cdot\|_{\phi,S_0}$ that

$$\|\psi\|_{\phi,S_0} \leq C_N \int_{\mathbb{R}^{2n}} (1+|z|)^{-N} dz,$$

and hence $\|\psi\|_{\phi,S_0} < \infty$ if we choose $N > 2n$. To prove the density of $\mathcal{S}(\mathbb{R}^n)$ in $S_0(\mathbb{R}^n)$, let us choose an exhaustive sequence (K_j) of compact subsets of $\mathbb{R}^{2n}$ (i. e., $K_j \subset \mathring{K}_{j+1}$ and $\mathbb{R}^{2n} = \bigcup_j K_j$) and set $\Psi_j = W(\psi,\phi)\chi_j$ where χ_j is a C^∞ function such that $0 \leq \chi_j \leq 1$, equal to one on K_j and supported in K_{j+1}. Recalling that we have by the reconstruction formula (3.17), and since $\|\phi\|_{L^2} = 1$,

$$\psi = 2^n \int_{\mathbb{R}^{2n}} W(\psi,\phi)(z)\widehat{R}(z)\phi dz,$$

we define an approximating function by

$$\psi_j = 2^n \int_{\mathbb{R}^{2n}} \Psi_j(z)\widehat{R}(z)\phi dz.$$

Since $W(\psi,\phi) \in \mathcal{S}(\mathbb{R}^{2n})$, it is easy to see, using Leibniz's differentiation rule for products, that $\psi_j \in \mathcal{S}(\mathbb{R}^n)$. We thus have

$$\psi - \psi_j = 2^n \int_{\mathbb{R}^{2n}} W(\psi,\phi)(z)(1-\chi_j(z))\widehat{R}(z)\phi dz. \tag{10.7}$$

Let us show

$$\lim_{j\to\infty} \|\psi - \psi_j\|_{\phi,S_0} = 0,$$

which will prove our assertion. By definition of the cross-Wigner transform, we have

$$W(\psi-\psi_j,\phi)(z') = \left(\frac{1}{\pi\hbar}\right)^n (\widehat{R}(z')(\psi-\psi_j)|\phi)_{L^2(\mathbb{R}^n)}$$

$$= \left(\frac{1}{\pi\hbar}\right)^n (\psi - \psi_j|\widehat{R}(-z')\phi)_{L^2(\mathbb{R}^n)}$$

where we have used the fact that $\widehat{R}(z')^* = \widehat{R}(-z')$; using (10.9), we get

$$W(\psi - \psi_j, \phi)(z') = \left(\frac{2}{\pi\hbar}\right)^n \times \int_{\mathbb{R}^{2n}} W(\psi, \phi)(z)(1 - \chi_j(z))(\widehat{R}(z)\phi|\widehat{R}(-z')\phi)_{L^2(\mathbb{R}^n)} dz.$$

Integrating with respect to z', we get, by definition (10.1) of the norm $\|\cdot\|_{\phi,S_0}$,

$$\|\psi - \psi_j\|_{\phi,S_0} = \|W(\psi - \psi_j, \phi)\|_{L^1(\mathbb{R}^{2n})} \leq \left(\frac{2}{\pi\hbar}\right)^n \int_{\mathbb{R}^{2n}} |W(\psi, \phi)(z)|(1 - \chi_j(z))|(\widehat{R}(z)\phi|\widehat{R}(-z')\phi)_{L^2(\mathbb{R}^n)}| dz.$$

By the Cauchy–Schwarz inequality, we have, since the operators $\widehat{R}(\cdot)$ are unitary and $\|\phi\|_{L^2} = 1$,

$$|(\widehat{R}(z)\phi|\widehat{R}(-z')\phi)_{L^2(\mathbb{R}^n)}| \leq 1,$$

so that

$$\|\psi - \psi_j\|_{\phi,S_0} \leq \left(\frac{2}{\pi\hbar}\right)^n \int_{\mathbb{R}^{2n}} |W(\psi, \phi)(z)|(1 - \chi_j(z)) dz.$$

Since

$$\int_{\mathbb{R}^{2n}} |W(\psi, \phi)(z)|(1 - \chi_j(z)) dz \leq \int_{\mathbb{R}^{2n}} |W(\psi, \phi)(z)| dz < \infty,$$

we conclude, using Lebesgue's dominated convergence theorem, that

$$\lim_{j\to\infty} \int_{\mathbb{R}^{2n}} |W(\psi, \phi)(z)|(1 - \chi_j(z)) dz = 0,$$

hence also $\lim_{j\to\infty} \|\psi - \psi_j\|_{\phi,S_0} = 0$. □

10.1.3 The Banach algebra property

We are going to see that the Feichtinger algebra is a Banach algebra; in addition, we will prove that this algebra is invariant under the action of the metaplectic group and that it enjoys a characteristic minimality property for the action of the Heisenberg–Weyl operators.

As we have seen that the topology defined on $S_0(\mathbb{R}^n)$ using the norm

$$\|\psi\|_{\phi,S_0} = \|W(\psi,\phi)\|_{L^1(\mathbb{R}^{2n})}$$

is independent of the choice of window ϕ. Let us prove that the normed space $S_0(\mathbb{R}^n)$ is complete.

Proposition 120. *The Feichtinger algebra $S_0(\mathbb{R}^n)$ is a Banach space for the norm $\|\cdot\|_{\phi,S_0}$.*

Proof. Let $(\psi_j)_j$ be a Cauchy sequence in $S_0(\mathbb{R}^n)$; then $(\Psi_j)_j = (W(\psi_j,\phi))_j$ is a Cauchy sequence in $L^1(\mathbb{R}^{2n})$. In fact,

$$W(\psi_j,\phi) - W(\psi_k,\phi) = W(\psi_j - \psi_k,\phi),$$

and hence, using Moyal's identity (3.13), we have

$$\|W(\psi_j,\phi) - W(\psi_k,\phi)\|_{L^2(\mathbb{R}^{2n})} = \left(\frac{1}{2\pi\hbar}\right)^n \|\psi_j - \psi_k\|_{L^2}\|\phi\|_{L^2},$$

proving that $W(\psi_j,\phi)$ is a Cauchy sequence if and only if $(\psi_j)_j$ is. The space $L^1(\mathbb{R}^{2n})$ being complete, there exists $\Psi \in L^1(\mathbb{R}^{2n})$ such that

$$\lim_{j\to\infty}\|\Psi - W(\psi_j,\phi)\|_{L^1(\mathbb{R}^{2n})} = 0.$$

Defining ψ by the formula

$$\psi(x) = \frac{2^n}{\|\phi\|_{L^2}} \int_{\mathbb{R}^{2n}} \Psi(z)\widehat{R}(z)\phi(x)dz, \tag{10.8}$$

one thereafter shows that $\psi \in S_0(\mathbb{R}^n)$ and that

$$\begin{aligned}\|\psi - \psi_j\|_{\phi,S_0} &= \lim_{j\to\infty}\|W(\psi - \psi_j,\phi)\|_{L^1(\mathbb{R}^{2n})}\\ &= \lim_{j\to\infty}\|\Psi - W(\psi_j,\phi)\|_{L^1(\mathbb{R}^{2n})} = 0,\end{aligned}$$

hence $S_0(\mathbb{R}^n)$ is complete as claimed. Let us show that $\mathcal{S}(\mathbb{R}^n) \subset S_0(\mathbb{R}^n)$. Let $\psi \in \mathcal{S}(\mathbb{R}^n)$; for every window ϕ, we have $W(\psi,\phi) \in \mathcal{S}(\mathbb{R}^{2n})$, hence for every $N > 0$ there exists $C_N > 0$ such that

$$|W(\psi,\phi)(z)| \le C_N(1 + |z|)^{-N}.$$

It follows, by definition of the norm $\|\cdot\|_{\phi,S_0}$, that

$$\|\psi\|_{\phi,S_0} \le C_N \int_{\mathbb{R}^{2n}} (1 + |z|)^{-N} dz,$$

and hence $\|\psi\|_{\phi,S_0} < \infty$ if we choose $N > 2n$. To prove the density of $\mathcal{S}(\mathbb{R}^n)$ in $S_0(\mathbb{R}^n)$, let us choose an exhaustive sequence (K_j) of compact subsets of $\mathbb{R}^{2n}$ (i. e., $K_j \subset \mathring{K}_{j+1}$ and $\mathbb{R}^{2n} = \bigcup_j K_j$) and set $\Psi_j = W(\psi,\phi)\chi_j$ where χ_j is a C^∞ function such that $0 \leq \chi_j \leq 1$, equal to one on K_j and supported in K_{j+1}. Recalling that we have, by the reconstruction formula (3.17), and since $\|\phi\|_{L^2} = 1$,

$$\psi = 2^n \int_{\mathbb{R}^{2n}} W(\psi,\phi)(z)\widehat{R}(z)\phi dz,$$

we define an approximating function by

$$\psi_j = 2^n \int_{\mathbb{R}^{2n}} \Psi_j(z)\widehat{R}(z)\phi dz.$$

Since $W(\psi,\phi) \in \mathcal{S}(\mathbb{R}^{2n})$, it is easy to see, using Leibniz's differentiation rule for products, that $\psi_j \in \mathcal{S}(\mathbb{R}^n)$. We thus have

$$\psi - \psi_j = 2^n \int_{\mathbb{R}^{2n}} W(\psi,\phi)(z)(1-\chi_j(z))\widehat{R}(z)\phi dz. \tag{10.9}$$

Let us show

$$\lim_{j\to\infty} \|\psi - \psi_j\|_{\phi,S_0} = 0,$$

which will prove our assertion. By definition of the cross-Wigner transform, we have

$$\begin{aligned} W(\psi-\psi_j,\phi)(z') &= \left(\frac{1}{\pi\hbar}\right)^n (\widehat{R}(z')(\psi-\psi_j)|\phi)_{L^2(\mathbb{R}^n)} \\ &= \left(\frac{1}{\pi\hbar}\right)^n (\psi-\psi_j|\widehat{R}(-z')\phi)_{L^2(\mathbb{R}^n)} \end{aligned}$$

where we have used the fact that $\widehat{R}(z')^* = \widehat{R}(-z')$; using (10.9), we get

$$W(\psi-\psi_j,\phi)(z') = \left(\frac{2}{\pi\hbar}\right)^n \int_{\mathbb{R}^{2n}} W(\psi,\phi)(z)(1-\chi_j(z))(\widehat{R}(z)\phi|\widehat{R}(-z')\phi)_{L^2(\mathbb{R}^n)} dz.$$

Integrating with respect to z', we get, by definition (10.1) of the norm $\|\cdot\|_{\phi,S_0}$,

$$\begin{aligned} \|\psi-\psi_j\|_{\phi,S_0} &= \|W(\psi-\psi_j,\phi)\|_{L^1(\mathbb{R}^{2n})} \\ &\leq \left(\frac{2}{\pi\hbar}\right)^n \int_{\mathbb{R}^{2n}} |W(\psi,\phi)(z)|(1-\chi_j(z))|(\widehat{R}(z)\phi|\widehat{R}(-z')\phi)_{L^2(\mathbb{R}^n)}| dz. \end{aligned}$$

By the Cauchy–Schwarz inequality, we have, since the operators $\widehat{R}(\cdot)$ are unitary and $\|\phi\|_{L^2} = 1$,

$$|(\widehat{R}(z)\phi|\widehat{R}(-z')\phi)_{L^2(\mathbb{R}^n)}| \leq 1$$

so that

$$\|\psi - \psi_j\|_{\phi,S_0} \leq \left(\frac{2}{\pi\hbar}\right)^n \int_{\mathbb{R}^{2n}} |W(\psi,\phi)(z)|(1-\chi_j(z))dz.$$

Since

$$\int_{\mathbb{R}^{2n}} |W(\psi,\phi)(z)|(1-\chi_j(z))dz \leq \int_{\mathbb{R}^{2n}} |W(\psi,\phi)(z)|dz < \infty,$$

we conclude, using Lebesgue's dominated convergence theorem, that

$$\lim_{j\to\infty} \int_{\mathbb{R}^{2n}} |W(\psi,\phi)(z)|(1-\chi_j(z))dz = 0,$$

hence also $\lim_{j\to\infty} \|\psi - \psi_j\|_{\phi,S_0} = 0$. □

We next prove a convolution result:

Proposition 121. *Suppose that $\psi \in L^1(\mathbb{R}^n)$ and $\psi' \in S_0(\mathbb{R}^n)$. Then $\psi * \psi' \in S_0(\mathbb{R}^n)$, and we have*

$$\|\psi * \psi'\|_{\phi,S_0} \leq \|\psi\|_{L^1} \|\psi'\|_{\phi,S_0} \tag{10.10}$$

for every window $\phi \in \mathcal{S}(\mathbb{R}^n)$. Thus, if $\psi \in L^1(\mathbb{R}^n)$ and $\psi' \in S_0$, then $\psi * \psi' \in S_0(\mathbb{R}^n)$:

$$L^1(\mathbb{R}^n) * S_0(\mathbb{R}^n) \subset S_0(\mathbb{R}^n).$$

Proof. Recall that the cross-Wigner transform is defined by

$$W(\psi,\phi)(z) = \left(\frac{1}{\pi\hbar}\right)^n (\widehat{R}(z)\psi|\phi)_{L^2}$$

where $\widehat{R}(z_0)$ is the Grossmann–Royer reflection operator. This yields the formula

$$W(\psi,\phi)(z_0) = \left(\frac{1}{\pi\hbar}\right)^n e^{-\frac{2i}{\hbar}p_0\cdot x_0} \int_{\mathbb{R}^n} \psi(2x_0 - x)\phi_{p_0}(x)dx$$

with $\phi_{p_0}(x) = e^{\frac{2i}{\hbar}p_0\cdot x}\phi(x)$, that is

$$W(\psi,\phi)(z_0) = \left(\frac{1}{\pi\hbar}\right)^n e^{-\frac{2i}{\hbar}p_0\cdot x_0} \psi * \phi_{p_0}(2x_0). \tag{10.11}$$

It follows, in particular, that

$$\|\psi\|_{\phi,S_0} = \left(\frac{1}{2\pi\hbar}\right)^n \int_{\mathbb{R}^n} \|\psi * \phi_{p_0}\|_{L^1} dp_0. \tag{10.12}$$

Formula (10.11) now shows that

$$W(\psi * \psi', \phi)(z_0) = \left(\frac{1}{\pi\hbar}\right)^n e^{-\frac{2i}{\hbar}p_0\cdot x_0}\psi * \psi' * \phi_{p_0}(2x_0),$$

and hence, by (10.12),

$$\|\psi * \psi'\|_{\phi,S_0} = \left(\frac{1}{2\pi\hbar}\right)^n \int_{\mathbb{R}^n} \|\psi * (\psi' * \phi_{p_0})\|_{L^1} dp_0.$$

Since $L^1(\mathbb{R}^n)$ is a convolution algebra, we have

$$\|\psi * (\psi' * \phi_{p_0})\|_{L^1} \leq \|\psi\|_{L^1}\|\psi' * \phi_{p_0}\|_{L^1},$$

and we obtain the inequality

$$\|\psi * \psi'\|_{\phi_0} \leq \left(\frac{1}{2\pi\hbar}\right)^n \|\psi\|_{L^1} \int_{\mathbb{R}^n} \|\psi' * \phi_{p_0}\|_{L^1} dp_0,$$

that is, using again (10.12),

$$\|\psi * \psi'\|_{\phi,S_0} \leq \|\psi\|_{L^1}\ \|\psi\|_{\phi,S_0}$$

which we set out to prove. □

Corollary 122. *The Banach space $S_0(\mathbb{R}^n)$ is an algebra for both pointwise multiplication and convolution: if ψ and ψ' are in $S_0(\mathbb{R}^n)$, then $\psi\psi' \in S_0(\mathbb{R}^n)$ and $\psi * \psi' \in S_0(\mathbb{R}^n)$.*

Proof. Since $\psi\psi'$ and $\psi*\psi'$ are interchangeable by the Fourier transform F, and $S_0(\mathbb{R}^n)$ being invariant under F in view of Proposition 123, (iii), it is sufficient to show that $\psi * \psi' \in S_0(\mathbb{R}^n)$ if $\psi \in S_0(\mathbb{R}^n)$ and $\psi' \in S_0(\mathbb{R}^n)$. This follows from the inequality (10.10) since $S_0(\mathbb{R}^n) \subset L^1(\mathbb{R}^n)$ in view of Proposition 118. □

10.2 The metaplectic invariance of $S_0(\mathbb{R}^n)$

Feichtinger's algebra is invariant (as are $L^2(\mathbb{R}^n)$ and $\mathcal{S}(\mathbb{R}^n)$) under the action of metaplectic operators and displacements:

Proposition 123. *Let $\psi \in S_0(\mathbb{R}^n)$, $\widehat{S} \in \mathrm{Mp}(n)$, and $z_0 \in \mathbb{R}^n$. We have:* (i) $\widehat{S}\psi \in S_0(\mathbb{R}^n)$; (ii) $\widehat{D}(z_0)\psi \in S_0(\mathbb{R}^n)$; (iii) *in particular, $\psi \in S_0(\mathbb{R}^n)$ if and only if $F\psi \in S_0(\mathbb{R}^n)$.*

Proof. We have $\psi \in S_0(\mathbb{R}^n)$ if and only if $W\psi \in L^1(\mathbb{R}^n)$. In view of the symplectic covariance property

$$W(\widehat{S}\psi)(z) = W\psi(S^{-1}z)$$

of the Wigner function, we have $W(\widehat{S}\psi)(z) = W\psi(S^{-1}z)$ where $S \in \mathrm{Sp}(2n, R)$ is the projection of $\widehat{S}$. Since $\det S = 1$, we have

$$\int_{\mathbb{R}^{2n}} |W\psi(S^{-1}z)|dz = \int_{\mathbb{R}^{2n}} |W(\widehat{S}\psi)(z)|dz,$$

hence $W(\widehat{S}\psi) \in L^1(\mathbb{R}^n)$ if and only if $W\psi \in L^1(\mathbb{R}^n)$. On the other hand, by formula (3.8),

$$W(\widehat{D}(z_0)\psi, \phi) = e^{\frac{i}{\hbar}\sigma(z,z_0)} W(\psi, \phi)\left(z - \frac{1}{2}z_0\right),$$

hence, for every window $\phi \in \mathcal{S}(\mathbb{R}^n)$,

$$\begin{aligned}\int_{\mathbb{R}^{2n}} |W(\widehat{T}(z_0)\psi, \phi)(z)|dz &= \int_{\mathbb{R}^{2n}} \left|W\psi\left(z - \frac{1}{2}z_0\right)\right|dz \\ &= \int_{\mathbb{R}^{2n}} |W\psi(z)|dz\end{aligned}$$

so that $W(\widehat{D}(z_0)\psi) \in S_0(\mathbb{R}^n)$ if and only if $W\psi \in L^1(\mathbb{R}^n)$. (iii) It follows from the fact that the Fourier transform F is related to the generator $\widehat{J}$ of $\mathrm{Mp}(n)$ by the formula $F = i^{n/2}\widehat{J}$. □

Corollary 124. *Every $\psi \in S_0(\mathbb{R}^n)$ is bounded, and we have* $\lim_{z\to\infty} \psi = 0$.

Proof. This follows from property (iii) in the previous proposition: Since ψ is continuous, it suffices to prove that $\lim_{z\to\infty} \psi = 0$. We have $F^{-1}\psi \in S_0(\mathbb{R}^n) \subset L^1(\mathbb{R}^n)$, hence $\psi = F(F^{-1}\psi)$ has limit 0 at infinity in view of the Riemann–Lebesgue lemma. □

10.3 The dual space $S_0'(\mathbb{R}^n)$

In this section, we will study the Banach Gelfand triple

$$(S_0(\mathbb{R}^n), L^2(\mathbb{R}^n), S_0'(\mathbb{R}^n))$$

where $S_0'(\mathbb{R}^n)$ is the dual space of the Feichtinger algebra $S_0(\mathbb{R}^n)$.

Let us denote by $S_0'(\mathbb{R}^n)$ the dual Banach space of $S_0(\mathbb{R}^n)$. It is the space of all bounded linear functionals on $S_0(\mathbb{R}^n)$. Since $S_0(\mathbb{R}^n)$ is the smallest Banach space isometrically invariant under the action of the affine metaplectic group (and hence under the Heisenberg–Weyl operators), its dual is essentially the largest space of distributions with this property.

The following result characterizes the distribution space $S_0'(\mathbb{R}^n)$:

Proposition 125. *The Banach space $S_0'(\mathbb{R}^n)$ consists of all $\psi \in S'(\mathbb{R}^n)$ such that $W(\psi,\phi) \in L^\infty(\mathbb{R}^{2n})$ for one (and hence all) window $\phi \in S_0(\mathbb{R}^n)$; the duality bracket is given by the pairing*

$$(\psi,\psi') = \int_{\mathbb{R}^{2n}} W(\psi,\phi)(z)\overline{W(\psi',\phi)(z)}dz, \tag{10.13}$$

and the formula

$$\|\psi\|_{\phi,S_0'(\mathbb{R}^n)} = \sup_{z\in\mathbb{R}^{2n}} |W(\psi,\phi)(z)| \tag{10.14}$$

defines a norm on $S_0'(\mathbb{R}^n)$.

Proof. We omit the proof of this result; see the comments and references at the end of the chapter. □

It readily follows from this characterization that:

Proposition 126. *The Dirac distribution δ is in $S_0'(\mathbb{R}^n)$; more generally, $\delta_a \in S_0'(\mathbb{R}^n)$ where $\delta_a(x) = \delta(x-a)$.*

Proof. By translation, it is sufficient to assume $a = 0$. We have, by definition of the cross-Wigner transform,

$$W(\delta,\phi)(z_0) = \left(\frac{1}{\pi\hbar}\right)^n (\widehat{R}(z_0)\delta|\overline{\phi})_{L^2}$$

and also

$$\widehat{R}(z_0)\delta(x) = e^{\frac{2i}{\hbar}p_0\cdot(x-x_0)}\delta(2x_0-x) = e^{\frac{2i}{\hbar}p_0\cdot x_0}\delta(2x_0-x).$$

It follows that

$$W(\delta,\phi)(z_0) = \left(\frac{1}{\pi\hbar}\right)^n e^{\frac{2i}{\hbar}p_0\cdot x_0}\overline{\phi}(2x_0),$$

and hence

$$|W(\delta,\phi)(z_0)| \le \left(\frac{1}{\pi\hbar}\right)^n \|\phi\|_\infty.$$

It follows that $\delta \in S_0'(\mathbb{R}^n)$. □

Recall that a Banach Gelfand triple is a triple $(\mathcal{B}, H, \mathcal{B}')$ consisting of a Banach space $\mathcal{B}$ that is continuously and densely embedded into a Hilbert space H, which in turn is w^*-continuously and densely embedded into the dual Banach space $\mathcal{B}'$. One identifies H with its dual H^*, and the scalar product on H thus extends in a natural way into a pairing between $\mathcal{B} \subset H$ and $\mathcal{B}' \supset H$. A typical example is the triple $(\mathcal{S}(\mathbb{R}^n), L^2(\mathbb{R}^n), \mathcal{S}'(\mathbb{R}^n))$.

Given a Gelfand triple $(\mathcal{B}, H, \mathcal{B}')$, every $\phi \in \mathcal{B}$ has an expansion with respect to the generalized eigenvectors ψ_α which generalizes the usual expansion with respect to a basis of eigenvectors. A classical example is the following: Consider the Gelfand triple $(S_0(\mathbb{R}^n), L^2(\mathbb{R}^n), S_0'(\mathbb{R}^n))$, and choose $\widehat{A} = -i\hbar\partial_{x_j}$. The generalized eigenvalues of $\widehat{A}$ are the functions $\chi_p(x) = e^{ip\cdot x/\hbar}$ $(p \in \mathbb{R}^n)$, and the corresponding expansion can be written as the Fourier inversion formula

$$\psi(x) = \left(\frac{1}{2\pi\hbar}\right)^{n/2} \int_{\mathbb{R}^n} e^{\frac{i}{\hbar}p\cdot x} F\psi(p)dp \tag{10.15}$$

(see Feichtinger et al. in [27] for a detailed discussion of the Fourier transform within the context of the Banach Gelfand triple $(S_0(\mathbb{R}^n), L^2(\mathbb{R}^n), S_0'(\mathbb{R}^n)))$.

An important feature of Gelfand triples is the existence of a kernel theorem, which is much more useful both for theoretical and practical purposes than the usual kernel theorem of Schwartz. We denote as usual by $((\cdot,\cdot)_{L^2})_{L^2}$ the distributional bracket for distributions on $\mathbb{R}^{2n}$.

Proposition 127. *The following properties hold:* (i) *Every linear bounded operator* $\widehat{A} : S_0(\mathbb{R}^n) \longrightarrow S_0'(\mathbb{R}^n)$ *has a kernel* $K \in S_0'(\mathbb{R}^n \times \mathbb{R}^n)$, *that is* $(A\psi, \phi)_{L^2} = ((K_A, \phi \otimes \psi)_{L^2})_{L^2}$ *for* ψ *and* ϕ *in* $S_0(\mathbb{R}^n)$. (ii) *Conversely, every* $K \in S_0'(\mathbb{R}^n \times \mathbb{R}^n)$ *defines by the previous formula a bounded operator* $S_0(\mathbb{R}^n) \longrightarrow S_0'(\mathbb{R}^n)$.

Formally we can thus write

$$A\psi(x) = \int_{\mathbb{R}^n} K(x,y)\psi(y)dy$$

for some $K \in S_0'(\mathbb{R}^n \times \mathbb{R}^n)$ when $A : S_0(\mathbb{R}^n) \longrightarrow S_0'(\mathbb{R}^n)$ is a continuous operator. This result was announced by Feichtinger in [25] and proven in [26]. See [40], § 11.4 for a detailed proof, comments, and various extensions.

10.4 The modulation spaces $M^1_s(\mathbb{R}^n)$

Closely related to the Feichtinger algebra are the modulation spaces $M^1_s(\mathbb{R}^n)$. Let us introduce the following notation. For $z \in \mathbb{R}^{2n}$, we set $\langle z \rangle = (1 + |z|^2)^{1/2}$. The function $z \longmapsto \langle z \rangle$ is the Weyl symbol of the pseudodifferential operator $(1 - \Delta)^{1/2}$ where Δ is the Laplacian in the z variables. For $s \geq 0$, we denote by $L^1_s(\mathbb{R}^{2n})$ the weighted L^1 space defined by

$$L_s^1(\mathbb{R}^{2n}) = \{\rho : \mathbb{R}^{2n} \longrightarrow \mathbb{C} : \langle\cdot\rangle^s \rho \in L^1(\mathbb{R}^{2n})\}.$$

Thus, a function ρ is in $L_s^1(\mathbb{R}^{2n})$ if and only if

$$\|\rho\|_{L_s^1(\mathbb{R}^{2n})} = \int_{\mathbb{R}^{2n}} |\rho(z)|\langle z\rangle^s dz < \infty.$$

$L_2^1\mathbb{R}^{2n}$ is clearly a complex vector space on which the mapping $\rho \longmapsto \|\rho\|_{L_2^1(\mathbb{R}^{2n})}$ is a norm; $L_s^1(\mathbb{R}^{2n})$ is complete for that norm and is thus a Banach space.

Definition 128. A function ψ belongs to the modulation space $M_s^1(\mathbb{R}^n)$ if and only if $W(\psi, \phi) \in L_s^1(\mathbb{R}^{2n})$ for every window $\phi \in \mathcal{S}(\mathbb{R}^n)$.

Of course, $M_0^1(\mathbb{R}^n) = S_0(\mathbb{R}^n)$, and more generally $M_s^1(\mathbb{R}^n) \subset S_0(\mathbb{R}^n)$. Hence, in view of the inclusions (10.5),

$$M_s^1(\mathbb{R}^n) \subset C^0(\mathbb{R}^n) \cap L^1(\mathbb{R}^n) \cap F(L^1(\mathbb{R}^n)).$$

Exactly as in the case of Feichtinger's algebra (Proposition 117), it suffices to check the definition for *one* window:

Proposition 129. (i) *We have $\psi \in M_s^1(\mathbb{R}^n)$ if and only if $W(\psi, \phi) \in L_s^1(\mathbb{R}^{2n})$ for one window ϕ.* (ii) *The mappings $\psi \longmapsto \|\psi\|_{\phi, M_s^1}$ defined by*

$$\|\psi\|_{\phi, M_s^1} = \|W(\psi, \phi)\|_{L_s^1(\mathbb{R}^{2n})} = \int_{\mathbb{R}^{2n}} |W(\psi, \phi)(z)|\langle z\rangle^s dz$$

form a family of equivalent norms, and the topology on $M_s^1(\mathbb{R}^n)$ thus defined makes it into a Banach space.

In fact, most of the properties we have proven for $S_0(\mathbb{R}^n)$ are also true for the modulation spaces $M_s^1(\mathbb{R}^n)$ and are proven quite similarly. The density result Proposition 119 for $S_0(\mathbb{R}^n)$ extends to $M_s^1(\mathbb{R}^n)$:

Proposition 130. *The Schwartz space $\mathcal{S}(\mathbb{R}^n)$ is a dense subspace of each of the modulation spaces $M_s^1(\mathbb{R}^n)$.*

We also have the metaplectic invariance property extending Proposition 123:

Proposition 131. *The modulation space $M_s^1(\mathbb{R}^n)$ is invariant under the action of the metaplectic group $\mathrm{Mp}(n)$: if $\widehat{S} \in \mathrm{Mp}(n)$, then $\widehat{S}\psi \in M_s^1(\mathbb{R}^n)$ if and only if $\psi \in M_s^1(\mathbb{R}^n)$. In particular, $M_s^1(\mathbb{R}^n)$ is invariant under the Fourier transform $\psi \in M_s^1(\mathbb{R}^n)$ if and only if $F\psi \in M_s^1(\mathbb{R}^n)$.*

The modulation spaces $M_s^1(\mathbb{R}^n)$ behave well under the action of the Heisenberg displacement operator:

Proposition 132. *Each space* $M_s^1(\mathbb{R}^n)$ *is invariant under the action of the Heisenberg displacement operator* $\widehat{D}(z)$, *and there exists a constant* $C > 0$ *such that*

$$\|\widehat{D}(z_0)\psi\|_{\phi,M_s^1} \leq C\langle z_0\rangle^s\|\psi\|_{\phi,M_s^1} \tag{10.16}$$

for every $z_0 \in \mathbb{R}^{2n}$.

Proof. The cross-Wigner transform satisfies

$$W(\widehat{T}(z_0)\psi,\phi)(z) = e^{-\frac{i}{\hbar}\sigma(z,z_0)}W(\psi,\phi)\left(z - \frac{1}{2}z_0\right)$$

(property (3.11)), hence it suffices to show that $L_s^1(\mathbb{R}^{2n})$ is invariant under phase space translations $T(z_0) : z \longmapsto z + z_0$. In view of the submultiplicative property $\langle z + z_0\rangle^s \leq \langle z\rangle^s\langle z_0\rangle^s$ of the weight function $\langle\cdot\rangle^s$, we have

$$\begin{aligned}\|T(z_0)\rho\|_{L_s^1(\mathbb{R}^{2n})} &= \int_{\mathbb{R}^{2n}} |\rho(z - z_0)|\langle z\rangle^s dz \\ &= \int_{\mathbb{R}^{2n}} |\rho(z)|\langle z + z_0\rangle^s dz \\ &\leq \langle z_0\rangle^s \int_{\mathbb{R}^{2n}} |\rho(z)|\langle z\rangle^s dz,\end{aligned}$$

hence (10.16) and $\widehat{D}(z)\psi \in M_s^1(\mathbb{R}^n)$. □

10.5 Comments and references

We recommend Jakobsen's recent review [48]; of the Feichtinger algebra. The most detailed reference is still Gröchenig's treatise [40].

The Feichtinger algebra $S_0(\mathbb{R}^n)$ and its dual $S_0'(\mathbb{R}^n)$ were introduced by Hans Feichtinger in the early 1980s. It is a particular case of the class of modulation spaces $M_v^{p,q}(\mathbb{R}^n)$ that play an important role in many theoretical and practical questions in analysis. Modulation spaces were originally designed to study phase-space concentration problems in time-frequency analysis. In the traditional approaches (see e. g. Gröchenig [40]), the Feichtinger algebra is defined not in terms of the cross-Wigner transform, as we do, but rather using the short-time Fourier transform (also called the Gabor transform)

$$V_\phi\psi(z) = \int_{\mathbb{R}^n} e^{-2\pi ip\cdot x'}\psi(x')\overline{\phi(x' - x)}dx'. \tag{10.17}$$

The latter is related to the cross-Wigner transform by the formula

$$W(\psi,\phi)(z) = \left(\frac{2}{\pi\hbar}\right)^{n/2} e^{\frac{2i}{\hbar}p\cdot x}V_{\phi^\vee_{\sqrt{2\pi\hbar}}}\psi_{\sqrt{2\pi\hbar}}(z\sqrt{\frac{2}{\pi\hbar}}) \tag{10.18}$$

where $\psi_{\sqrt{2\pi\hbar}}(x) = \psi(x\sqrt{2\pi\hbar})$ and $\phi^{\vee}(x) = \phi(-x)$.

We have seen (Proposition 123) that Feichtinger's algebra is invariant under displacements and metaplectic operators. It is in fact the smallest Banach algebra in $\mathcal{S}'(\mathbb{R}^n)$ having this property; see Gröchenig [40]. For a proof of Proposition 125 see [40], §11.3.

The use of the Gelfand triple $(S_0(\mathbb{R}^n), L^2(\mathbb{R}^n), S_0'(\mathbb{R}^n))$ not only offers a better description of self-adjoint operators, but it also makes possible simplification of many proofs. Here is a typical situation. Given a Gelfand triple $(\mathcal{B}, H, \mathcal{B}')$, one proves that every self-adjoint operator $A : \mathcal{B} \longrightarrow \mathcal{B}$ has a complete family of generalized eigenvalues $(\psi_\alpha)_\alpha = \{\psi_\alpha \in \mathcal{B}' : \alpha \in \mathbb{A}\}$ ($\mathbb{A}$ an index set), defined as follows: For every $\alpha \in \mathbb{A}$, there exists $\lambda_\alpha \in \mathbb{C}$ such that

$$(\psi_\alpha | A\phi) = \lambda_\alpha(\psi_\alpha|\phi) \quad \text{for every } \phi \in \mathcal{B}.$$

A basic example, in the case $n = 1$, is the operator $\widehat{x}$ of multiplication by x. This operator has no eigenfunctions in $L^2(\mathbb{R})$, but since $\widehat{x}\delta_a(x) = x\delta(x-a) = a\delta_a(x)$, every $a \in \mathbb{R}$ is a generalized eigenvalue (with associated eigenfunction $\delta_a \in S_0'(\mathbb{R})$).

11 Hilbert–Schmidt operators

Hilbert–Schmidt operators play an essential role in harmonic and functional analysis. One of the reasons is that every density operator (and, more generally, every trace class operator) is the product of two Hilbert–Schmidt operators. The presentation in this chapter is quite conventional: We begin by defining and studying Hilbert–Schmidt operator on $L^2(\mathbb{R}^n)$, which are the operators with square integrable distributional kernels.

11.1 Hilbert–Schmidt operators on $L^2(\mathbb{R}^n)$

11.1.1 First definition

We begin by giving a simple working definition, which is actually sufficient for most of our needs since we will be, in practice, dealing almost exclusively with the Hilbert space $L^2(\mathbb{R}^n)$ in our study of quantum states in the forthcoming chapters.

Definition 133. A continuous operator $\widehat{A} : \mathcal{S}(\mathbb{R}^n) \longrightarrow \mathcal{S}'(\mathbb{R}^n)$ is called a Hilbert–Schmidt operator (or an integral Hilbert–Schmidt operator) if its kernel K is square integrable, that is, if there exists $K \in L^2(\mathbb{R}^n \times \mathbb{R}^n)$ such that we have

$$\widehat{A}\psi(x) = \int_{\mathbb{R}^n} K(x,y)\psi(y)dy \tag{11.1}$$

for all $\psi \in \mathcal{S}(\mathbb{R}^n)$. We denote by $\mathcal{L}_2(L^2(\mathbb{R}^n))$ the complex vector space of all Hilbert–Schmidt operators on $L^2(\mathbb{R}^n)$.

Remark 134. This definition can be extended to any measure space.

That $\mathcal{L}_2(L^2(\mathbb{R}^n))$ is a vector space is clear: If $\widehat{A}$ has a kernel K and $\widehat{B}$ a kernel L, then the kernel of $\widehat{A}+\widehat{B}$ is simply $K+L$, which is square integrable if K and L are. Similarly, for $\lambda \in \mathbb{C}$, the kernel of $\lambda\widehat{A}$ is λK which is square integrable if and only if K is. It is easy to check, using formula (11.1), that if $\widehat{A}$ is a Hilbert–Schmidt operator, then so is its adjoint $\widehat{A}^*$ and

$$\widehat{A}^*\psi(x) = \int_{\mathbb{R}^n} \overline{K(y,x)}\psi(y)dy. \tag{11.2}$$

A Hilbert–Schmidt operator $\widehat{A}$ can be uniquely extended into a continuous bounded operator $L^2(\mathbb{R}^n) \longrightarrow L^2(\mathbb{R}^n)$, also denoted $\widehat{A}$: For $\psi \in \mathcal{S}(\mathbb{R}^n)$, we have, by the Cauchy–Schwarz inequality,

$$|\widehat{A}\psi(x)|^2 \leq \int_{\mathbb{R}^n} |K(x,y)|^2 dy \int_{\mathbb{R}^n} |\psi(y)|^2 dy,$$

https://doi.org/10.1515/9783110722772-011

and hence, integrating with respect to x,

$$\|\widehat{A}\psi\|_{L^2}^2 \leq \|K\|_{L^2(\mathbb{R}^{2n})}^2 \|\psi\|_{L^2}^2. \tag{11.3}$$

This inequality extends to all $\psi \in L^2(\mathbb{R}^n)$ by density and continuity, hence $\widehat{A}$ is a continuous operator $L^2(\mathbb{R}^n) \longrightarrow L^2(\mathbb{R}^n)$. It also shows that the operator norm $\|\widehat{A}\|$ of a Hilbert–Schmidt operator $\widehat{A}$ satisfies

$$\|\widehat{A}\| = \sup_{\|\psi\|_{L^2}=1} \|\widehat{A}\psi\|_{L^2} \leq \|K\|_{L^2(\mathbb{R}^{2n})}. \tag{11.4}$$

The previous inequality motivates the following definition:

Definition 135. The *Hilbert–Schmidt norm* $\|\cdot\|_2$ on $\mathcal{L}_2(L^2(\mathbb{R}^n))$ is defined by

$$\|\widehat{A}\|_2 = \|K\|_{L^2(\mathbb{R}^{2n})} = \left(\int_{\mathbb{R}^{2n}} |K(x,y)|^2 dxdy \right)^{1/2} \tag{11.5}$$

where K is the (distributional) kernel of $\widehat{A} \in \mathcal{L}_2(L^2(\mathbb{R}^n))$.

The verification that $\|\cdot\|_2$ is indeed a norm is a trivial exercise left to the reader.

11.1.2 Second definition

An alternative definition of Hilbert–Schmidt operators (which can be extended to any separable Hilbert space) is the following:

Definition 136. An operator $\widehat{A} \in \mathcal{B}(L^2(\mathbb{R}^n))$ is called a Hilbert–Schmidt operator if there exists an orthonormal basis $(\psi_j)_j$ of $L^2(\mathbb{R}^n)$ such that

$$\sum_j \|\widehat{A}\psi_j\|_{L^2}^2 < \infty. \tag{11.6}$$

Equivalently,

$$\sum_j (\widehat{A}^*\widehat{A}\psi_j|\psi_j)_{L^2} < \infty. \tag{11.7}$$

The choice of basis $(\psi_j)_j$ for which (11.6) holds is irrelevant: If the condition (11.6) holds for one orthonormal basis, then it holds for all, and the sum $\sum_j \|\widehat{A}\psi_j\|_{L^2}^2$ does moreover not depend on the choice of basis. Let us prove this essential property.

Proposition 137. *Let $\widehat{A} \in \mathcal{L}_2(L^2(\mathbb{R}^n))$. Let $(\psi_j)_j$ and $(\phi_j)_j$ be orthonormal bases of $L^2(\mathbb{R}^n)$, and assume that $\sum_j \|\widehat{A}\psi_j\| < \infty$. Then*

$$\sum_j \|\widehat{A}\phi_j\|_{L^2}^2 = \sum_j \|\widehat{A}\psi_j\|^2 < \infty.$$

Proof. The Parseval identity

$$\|\psi\|_{L^2}^2 = \sum_k |(\psi|\phi_k)|^2 \tag{11.8}$$

implies that

$$\|\widehat{A}\psi_j\|_{L^2}^2 = \sum_k |(\widehat{A}\psi_j|\phi_k)_{L^2}|^2.$$

Summing over j, we get

$$\begin{aligned}\sum_j \|\widehat{A}\psi_j\|_{L^2}^2 &= \sum_{j,k} |(\widehat{A}\psi_j|\phi_k)_{L^2}|^2 \\ &= \sum_{j,k} |(\psi_j|\widehat{A}^*\phi_k)_{L^2}|^2.\end{aligned}$$

Applying again the Parseval identity, we have

$$\sum_j |(\psi_j|\widehat{A}^*\phi_k)_{L^2}|^2 = \|\widehat{A}^*\phi_k\|_{L^2}^2,$$

and hence

$$\sum_j \|\widehat{A}\psi_j\|_{L^2}^2 = \sum_k \|\widehat{A}^*\phi_k\|_{L^2}^2 < \infty.$$

Taking $(\phi_j)_j = (\psi_j)_j$, we have $\sum_k \|\widehat{A}^*\psi_k\|_{L^2}^2 < \infty$, hence the adjoint of $\widehat{A}$ is also a Hilbert–Schmidt operator; we may thus replace $\widehat{A}$ by $\widehat{A}^*$ in the previous formula, which yields $\sum_k \|\widehat{A}\phi_k\|_{L^2}^2 < \infty$ as claimed. □

Remark 138. In the course of the previous proof, we have shown that $\widehat{A} \in \mathcal{B}(L^2(\mathbb{R}^n))$ is a Hilbert–Schmidt operator if and only if its adjoint $\widehat{A}^*$ is — which is already clear if one uses Definition 11.1 since the kernel of the adjoint is $(x,y) \longmapsto \overline{K(y,x)}$ which is square integrable if and only if $(x,y) \longmapsto K(x,y)$ is.

11.1.3 Equivalence of the definitions; the Hilbert–Schmidt norm

There remains to prove that both definitions (11.1) and (11.6) are equivalent. This will enable us to give an alternative definition of the Hilbert–Schmidt norm.

Proposition 139. (i) *An operator $\widehat{A}$ on $L^2(\mathbb{R}^n)$ satisfies $\sum_j \|\widehat{A}\psi_j\|_{L^2}^2 < \infty$ for one (and hence every) orthonormal basis $(\psi_j)_j$ of $L^2(\mathbb{R}^n)$ if and only if it has a kernel $K \in L^2(\mathbb{R}^n \times \mathbb{R}^n)$.* (ii) *The Hilbert–Schmidt norm of $\widehat{A} \in \mathcal{L}_2(L^2(\mathbb{R}^n))$ is given by*

$$\|\widehat{A}\|_2 = \left(\sum_j \|\widehat{A}\psi_j\|_{L^2}^2\right)^{1/2} = \left(\sum_j (\widehat{A}^*\widehat{A}\psi_j|\psi_j)_{L^2}\right)^{1/2}. \tag{11.9}$$

(iii) *The absolutely convergent series*

$$(\widehat{A}|\widehat{B})_2 = \sum_j (\widehat{A}\psi_j|\widehat{B}\psi_j)_{L^2}$$

defines a scalar product on $\mathcal{L}_2(L^2(\mathbb{R}^n))$ *such that* $\|\widehat{A}\|_2 = ((\widehat{A}|\widehat{A})_2)^{1/2}$, *and* $\mathcal{L}_2(L^2(\mathbb{R}^n))$ *is a Hilbert space for this scalar product.*

Proof. (i) Let $\widehat{A}$ be a Hilbert–Schmidt operator on $L^2(\mathbb{R}^n)$ in the sense of definition (11.6), and choose an orthonormal basis $(\psi_j)_j$ of $L^2(\mathbb{R}^n)$. The family $(\psi_i \otimes \psi_j)_{i,j}$ of tensor products is an orthonormal basis of $L^2(\mathbb{R}^n \times \mathbb{R}^n)$. Consider the kernel

$$K(x,y) = \sum_{i,j} (\widehat{A}\psi_i|\psi_j)\psi_j(x)\overline{\psi_i(y)}. \tag{11.10}$$

We have

$$\begin{aligned}\int_{\mathbb{R}^{2n}} |K(x,y)|^2 dxdy &\leq \sum_{i,j} |(\widehat{A}\psi_i|\psi_j)_{L^2}|^2 \|\psi_j \otimes \overline{\psi_i}\|^2_{L^2(\mathbb{R}^{2n})} \\ &= \sum_{i,j} |(\widehat{A}\psi_i|\psi_j)_{L^2}|^2 < \infty,\end{aligned}$$

hence $K \in L^2(\mathbb{R}^n \times \mathbb{R}^n)$. By definition (11.10) of K, we have

$$\begin{aligned}\int_{\mathbb{R}^n} K(x,y)\psi(y)dy &= \sum_{i,j} (\widehat{A}\psi_i|\psi_j)\psi_j(x) \int_{\mathbb{R}^n} \psi(y)\overline{\psi_i(y)}dy \\ &= \sum_{i,j} (\widehat{A}\psi_i|\psi_j)_{L^2}\psi_j(x)(\psi|\psi_i).\end{aligned}$$

Since $\psi = \sum_i (\psi|\psi_i)_{L^2} e_i$ and $\widehat{A}\psi_i = \sum_j (\widehat{A}\psi_j|\psi_j)_{L^2} e_j$, we also have

$$\begin{aligned}\widehat{A}\psi(x) &= \sum_i (\psi|\psi_i)_{L^2} \widehat{A}\psi_i(x) \\ &= \sum_{i,j} (\psi|\psi_i)_{L^2} (\widehat{A}\psi_j|\psi_j)_{L^2} \psi_j(x),\end{aligned}$$

and hence

$$\widehat{A}\psi(x) = \int_{\mathbb{R}^n} K(x,y)\psi(y)dy. \tag{11.11}$$

Assume conversely that the kernel K of $\widehat{A} \in \mathcal{B}(L^2(\mathbb{R}^n))$ belongs to $L^2(\mathbb{R}^n \times \mathbb{R}^n)$. We can then find complex numbers c_{ij} such that $\sum_{i,j} |c_{ij}|^2 < \infty$ and

$$K(x,y) = \sum_{i,j} c_{ij}\psi_j(x) \otimes \overline{\psi_i(y)}. \tag{11.12}$$

The operator $\widehat{A}$ defined by

$$\widehat{A}\psi = \int_{\mathbb{R}^n} K(x,y)\psi(y)dy$$

is thus given by the absolutely convergent integral

$$\widehat{A}\psi = \sum_{i,j} c_{ij}(\psi|\psi_i)_{L^2}\psi_j.$$

Applying $\widehat{A}$ to the basis vectors ψ_k, we get, in particular, since the vectors ψ_j are orthonormal,

$$\widehat{A}\psi_k = \sum_{i,j} c_{ij}(\psi_k|\psi_i)_{L^2}\psi_j = \sum_j c_{kj}\psi_j,$$

and hence

$$\|\widehat{A}\psi_k\|_{L^2}^2 = (\widehat{A}\psi_k|\widehat{A}\psi_k)_{L^2} = \sum_j |c_{kj}|^2$$

so that

$$\sum_k \|\widehat{A}\psi_k\|_{L^2}^2 = \sum_{j,k} |c_{kj}|^2 < \infty, \tag{11.13}$$

and $\widehat{A}$ is thus a Hilbert–Schmidt operator. (ii) We have to show that

$$\int_{\mathbb{R}^{2n}} |K(x,y)|^2 dxdy = \sum_j \|\widehat{A}\psi_j\|_{L^2}^2.$$

In view of the identities (11.12) and (11.13), we have

$$\int_{\mathbb{R}^{2n}} |K(x,y)|^2 dxdy = \sum_{i,j} |c_{ij}|^2 = \sum_k \|\widehat{A}\psi_k\|_{L^2}^2.$$

(iii) We must show that $\sum_j |(\widehat{A}\psi_j|\widehat{B}\psi_j)_{L^2}| < \infty$ and is independent of the choice of orthonormal basis $(\psi_j)_j$. In view of the Cauchy–Schwarz inequality, we have

$$\begin{aligned}\sum_j |(\widehat{A}\psi_j|\widehat{B}\psi_j)_{L^2}| &\leq \sum_j \|\widehat{A}\psi_j\|_{L^2}\|\widehat{B}\psi_j\|_{L^2} \\ &\leq \frac{1}{2}\sum_j \|\widehat{A}\psi_j\|_{L^2}^2 + \|\widehat{B}\psi_j\|_{L^2}^2 < \infty\end{aligned}$$

hence the first assertion. That $(\widehat{A}|\widehat{B})_2$ is independent of the choice of orthonormal basis follows from the observation that, since $\|\widehat{A}\|_2 = ((\widehat{A}|\widehat{A})_2)^{1/2}$, we must have (Jordan and von Neumann's theorem)

$$\|\widehat{A} + \widehat{B}\|_2^2 + \|\widehat{A} - \widehat{B}\|_2^2 = 2(\|\widehat{A}\|_2^2 + \|\widehat{B}\|_2^2),$$

and the inner product $(\widehat{A}|\widehat{B})_2$ is uniquely determined by this requirement since it implies that

$$2(\widehat{A}|\widehat{B})_2 = \|\widehat{A}\|_2^2 + \|\widehat{B}\|_2^2 - \|\widehat{A} - \widehat{B}\|_2^2.$$

The proof of the completeness of $\mathcal{L}_2(L^2(\mathbb{R}^n))$ is omitted. □

11.2 Further properties of the Hilbert–Schmidt operators

11.2.1 Compactness of Hilbert–Schmidt operators

Let $A \in \mathcal{B}(L^2(\mathbb{R}^n))$. The absolute value $|\widehat{A}|$ of $\widehat{A}$ is defined by $|\widehat{A}| = (\widehat{A}^*\widehat{A})^{1/2}$. Let us show that the absolute value of a Hilbert–Schmidt operator is also a Hilbert–Schmidt operator and that the square root of a nonnegative Hilbert–Schmidt operator is also a Hilbert–Schmidt.

Proposition 140. *We have $\widehat{A} \in \mathcal{L}_2(L^2(\mathbb{R}^n))$ if and only if $|\widehat{A}| \in \mathcal{L}_2(L^2(\mathbb{R}^n))$.*

Proof. Let $(\psi_j)_j$ be an orthonormal basis of $L^2(\mathbb{R}^n)$. Assume that $\widehat{A} \in \mathcal{L}_2(L^2(\mathbb{R}^n))$. We have

$$\begin{aligned}\sum_j \||\widehat{A}|\psi_j\|_{L^2}^2 &= \sum_j ((\widehat{A}^*\widehat{A})^{1/2}\psi_j|(\widehat{A}^*\widehat{A})^{1/2}\psi_j)_{L^2} \\ &= \sum_j (\widehat{A}^*\widehat{A}\psi_j|\psi_j)_{L^2} \\ &= \sum_j (\widehat{A}\psi_j|\widehat{A}\psi_j)_{L^2} < \infty\end{aligned}$$

in view of definition (11.6). The sequence of these equalities can be reversed, so we conclude that, if $|\widehat{A}| \in \mathcal{L}_2(L^2(\mathbb{R}^n))$, then we also have $\widehat{A} \in \mathcal{L}_2(L^2(\mathbb{R}^n))$. □

Hilbert–Schmidt operators are bounded. They are in fact compact operators:

Proposition 141. *Every $\widehat{A} \in \mathcal{L}_2(L^2(\mathbb{R}^n))$ is a compact operator on $L^2(\mathbb{R}^n)$.*

Proof. It suffices to show that $\widehat{A}$ is the limit (in the operator norm) of a sequence of finite rank operators, and this will prove the result since the limit (in the operator norm) of a sequence of compact operators is compact and finite-rank operators are compact. Let $(\psi_j)_j$ be an orthonormal basis of $L^2(\mathbb{R}^n)$. Define now an operator $\widehat{A}_N$ on $L^2(\mathbb{R}^n)$ by

$$\widehat{A}_N\psi = \sum_{j=1}^{N} (\widehat{A}\psi|\psi_j)_{L^2}\psi_j.$$

Noting that $\widehat{A}\psi$ is given by the absolutely convergent series

$$\widehat{A}\psi = \sum_{j=1}^{\infty}(\psi|\psi_j)_{L^2}\widehat{A}\psi_j,$$

we have

$$(\widehat{A} - \widehat{A}_N)\psi = \sum_{j>N}(\psi|\psi_j)_{L^2}\widehat{A}\psi_j,$$

and hence,

$$\begin{aligned}\|(\widehat{A} - \widehat{A}_N)\psi\|_{L^2}^2 &\leq \sum_{j>N}|(\psi|\psi_j)_{L^2}|^2 \sum_{j>N}\|\widehat{A}\psi_j\|_{L^2}^2 \\ &\leq \|\psi\|_{L^2}\sum_{j>N}\|\widehat{A}\psi_j\|_{L^2}^2.\end{aligned}$$

This implies the inequality

$$\|\widehat{A} - \widehat{A}_N\|^2 \leq \sum_{j>N}\|\widehat{A}\psi_j\|_{L^2}^2;$$

the series $\sum_j \|\widehat{A}\psi_j\|_{L^2}^2$ being convergent (its limit is the square of the Hilbert–Schmidt norm $\|\widehat{A}\|_2$), this implies that $\lim_{N\to\infty}\|(\widehat{A} - \widehat{A}_N)\| = 0$. □

11.2.2 Ideal property

We are going to show that $\mathcal{L}_2(L^2(\mathbb{R}^n))$ is a left and right $\mathcal{B}(L^2(\mathbb{R}^n))$-module, in fact a two-sided ideal in $\mathcal{B}(L^2(\mathbb{R}^n))$.

Proposition 142. *The vector space $\mathcal{L}_2(L^2(\mathbb{R}^n))$ is a two-sided ideal in the algebra $\mathcal{B}(L^2(\mathbb{R}^n))$ of bounded operators on $L^2(\mathbb{R}^n)$: If $\widehat{A} \in \mathcal{L}_2(L^2(\mathbb{R}^n))$ and $\widehat{B} \in \mathcal{B}(L^2(\mathbb{R}^n))$, then $\widehat{A}\widehat{B}$ and $\widehat{B}\widehat{A}$ are both in $\mathcal{L}_2(L^2(\mathbb{R}^n))$, and we have*

$$\|\widehat{A}\widehat{B}\|_2 \leq \|\widehat{A}\|_2\,\|\widehat{B}\|, \quad \|\widehat{B}\widehat{A}\|_2 \leq \|\widehat{B}\|\|\widehat{A}\|_2 \tag{11.14}$$

where $\|\widehat{B}\|$ is the operator norm of $\widehat{B}$. In particular, $\mathcal{L}_2(L^2(\mathbb{R}^n))$ is a normed algebra.

Proof. Since Hilbert–Schmidt operators are compact, it is clear that $\widehat{A}\widehat{B}$ and $\widehat{B}\widehat{A}$ are both in $\mathcal{B}(L^2(\mathbb{R}^n))$. Let us show that they are in $\mathcal{L}_2(L^2(\mathbb{R}^n))$. For this, it suffices to prove the inequalities (11.14). Let $(\psi_j)_j$ be an orthonormal basis of $L^2(\mathbb{R}^n)$; we have $\|\widehat{B}\widehat{A}\psi_j\|_{L^2} \leq \|\widehat{B}\|\|\widehat{A}\psi_j\|_{L^2}$; summing over j yields $\|\widehat{B}\widehat{A}\|_2 \leq \|\widehat{B}\|\|\widehat{A}\|_2$. The inequality $\|\widehat{A}\widehat{B}\|_2 \leq \|\widehat{A}\|_2\|\widehat{B}\|$ follows, writing $\widehat{A}\widehat{B} = (\widehat{B}^*\widehat{A}^*)^*$. □

11.3 Hilbert–Schmidt and Weyl operators

The Schwartz kernel theorem tells us that every continuous operator $\widehat{A} : \mathcal{S}(\mathbb{R}^n) \longrightarrow \mathcal{S}'(\mathbb{R}^n)$ can be represented in the weak sense as $\langle \widehat{A}\psi, \phi\rangle = \langle K, \phi \otimes \psi\rangle$ where $K \in \mathcal{S}'(\mathbb{R}^n \times \mathbb{R}^n)$. Hilbert–Schmidt operators are, by definition, exactly those operators that are represented by a kernel $K \in L^2(\mathbb{R}^n \times \mathbb{R}^n)$. We also know that the relationship between the kernel of an operator and its Weyl symbol is given by the Fourier transform (formula (5.10))

$$a(x,p) = \int_{\mathbb{R}^n} e^{-\frac{i}{\hbar}p\cdot y} K\left(x + \frac{1}{2}y, x - \frac{1}{2}y\right) dy \tag{11.15}$$

or, equivalently,

$$K\left(x + \frac{1}{2}y, x - \frac{1}{2}y\right) = \left(\frac{1}{2\pi\hbar}\right)^n \int_{\mathbb{R}^n} e^{\frac{i}{\hbar}p\cdot y} a(x,p)dp. \tag{11.16}$$

Proposition 143. *Assume that* $\widehat{A} = \mathrm{Op_W}(a)$ *is a Hilbert–Schmidt operator. Then* $a \in L^2(\mathbb{R}^{2n})$ *and* $a_\sigma \in L^2(\mathbb{R}^{2n})$*; moreover:*

$$\|a\|_{L^2(\mathbb{R}^{2n})} = \|a_\sigma\|_{L^2(\mathbb{R}^{2n})} = (2\pi\hbar)^{n/2} \|K\|_{L^2(\mathbb{R}^n \times \mathbb{R}^n)}, \tag{11.17}$$

that is

$$\|\widehat{A}\|_2 = (2\pi\hbar)^{-n/2} \|a\|_{L^2(\mathbb{R}^{2n})}. \tag{11.18}$$

Proof. The equality $\|a\|_{L^2(\mathbb{R}^{2n})} = \|a_\sigma\|_{L^2(\mathbb{R}^{2n})}$ is obvious since the symplectic Fourier transform is unitary. Let us prove that $\|a\|_{L^2(\mathbb{R}^{2n})} = (2\pi\hbar)^{n/2}\|K\|_{L^2}$ when $K \in \mathcal{S}(\mathbb{R}^n \times \mathbb{R}^n)$; the proposition will follow using the density of $\mathcal{S}(\mathbb{R}^n \times \mathbb{R}^n)$ in $L^2(\mathbb{R}^n \times \mathbb{R}^n)$. In view of formula (11.15), the symbol a is, for fixed x, $(2\pi\hbar)^{n/2}$ times the Fourier transform of the function $y \longmapsto K(x + \frac{1}{2}y, x - \frac{1}{2}y)$, hence, by Plancherel's formula,

$$\int_{\mathbb{R}^n} |a(x,p)|^2 dp = (2\pi\hbar)^n \int_{\mathbb{R}^n} \left|K\left(x + \frac{1}{2}y, x - \frac{1}{2}y\right)\right|^2 dy \tag{11.19}$$

and, integrating with respect to x,

$$\begin{aligned}\int_{\mathbb{R}^{2n}} |a(z)|^2 dz &= (2\pi\hbar)^n \int_{\mathbb{R}^n}\left(\int_{\mathbb{R}^n} \left|K\left(x + \frac{1}{2}y, x - \frac{1}{2}y\right)\right|^2 dy\right)dx \\ &= (2\pi\hbar)^n \int_{\mathbb{R}^n \times \mathbb{R}^n} \left|K\left(x + \frac{1}{2}y, x - \frac{1}{2}y\right)\right|^2 dxdy\end{aligned}$$

where we have applied Fubini's theorem (the integrals are absolutely convergent since $(x,y) \longmapsto K(x+\frac{1}{2}y, x-\frac{1}{2}y)$ is in $\mathcal{S}(\mathbb{R}^n \times \mathbb{R}^n)$ because K is). Now set $x' = x + \frac{1}{2}y$ and $y' = x - \frac{1}{2}y$; we have $dx'dy' = dxdy$, hence

$$\int_{\mathbb{R}^{2n}} |a(z)|^2 dz = (2\pi\hbar)^n \int_{\mathbb{R}^n \times \mathbb{R}^n} |K(x',y')|^2 dx'dy'$$

which we set out to prove. Formula (11.18) follows by the definition of the Hilbert–Schmidt norm (Definition 135). □

11.4 Comments and references

The theory of Hilbert–Schmidt (and the related trace class operators) is a well-established topic in functional analysis. Their theory can be found in almost any introductory text to functional analysis. A good reference at the undergraduate level is MacCluer [57], another is the monograph by Blanchard and Brüning [6]. In our presentation, we have deliberately emphasized the operators acting on the square-integrable functions, thus privileging the model Hilbert space $L^2(\mathbb{R}^n)$. We have however always given alternative definitions in terms of orthonormal bases that can be immediately extended to arbitrary (separable) Hilbert spaces.

12 The trace class

Trace class operators are essential for the study of mixed quantum states and their density operators.

12.1 Definitions of trace class operators

12.1.1 Statement of the definitions

Let us give two definitions of trace class operators; we will thereafter show that they are equivalent. The first definition expresses these operators directly in terms of Hilbert–Schmidt operators:

Definition 144. The space $\mathcal{L}_1(L^2(\mathbb{R}^n))$ of trace class operators on $L^2(\mathbb{R}^n)$ consists of all operators $\widehat{A} \in \mathcal{B}(L^2(\mathbb{R}^n))$ that are the product of two Hilbert–Schmidt operators on $L^2(\mathbb{R}^n)$. That is, $\widehat{A} \in \mathcal{L}_1(L^2(\mathbb{R}^n))$ if and only if there exist $\widehat{B}$ and $\widehat{C}$ in $\mathcal{L}_2(L^2(\mathbb{R}^n))$ such that $\widehat{A} = \widehat{C}^*\widehat{B}$.

The notation $\widehat{A} = \widehat{C}^*\widehat{B}$ instead of, say, $\widehat{A} = \widehat{B}\widehat{C}$, might seem cumbersome: It will however make many arguments clearer as we will soon see. Note that this definition immediately implies the inclusion

$$\mathcal{L}_1(L^2(\mathbb{R}^n)) \subset \mathcal{L}_2(L^2(\mathbb{R}^n))$$

in view of the algebra property of $\mathcal{L}_2(L^2(\mathbb{R}^n))$ (Proposition 141). The compactness of trace class operators follows, so we have the following spectral decomposition result:

Proposition 145. *Let $\widehat{A}$ be a self-adjoint trace class operator on $L^2(\mathbb{R}^n)$. Then:* (i) *The set (λ_j) of eigenvalues λ_j of $\widehat{A}$ is at most countable and consists of non-negative numbers $\lambda_1 \geq \lambda_2 \geq \cdots \geq \lambda_j \geq \cdots \geq 0$ such that $\sum_j \lambda_j = 1$, and we have $\lim_{j\to\infty} \lambda_j = 0$ if (λ_j) is infinite.*

Our second definition is more in the spirit of that of Hilbert–Schmidt operators because it uses orthonormal bases in $L^2(\mathbb{R}^n)$:

Definition 146. A bounded linear operator $\widehat{A}$ on $L^2(\mathbb{R}^n)$ is of trace class if and only if there exists a pair of orthonormal bases $(\psi_j)_j$ and $(\phi_k)_k$ of $L^2(\mathbb{R}^n)$ such that

$$\sum_{j,k} (\widehat{A}\psi_j|\phi_k)_{L^2} < \infty \tag{12.1}$$

where the series is absolutely convergent.

It immediately follows from this definition that $\mathcal{L}_1(L^2(\mathbb{R}^n))$ is a vector space: If $\sum_j |(\widehat{A}_1\phi_j|\chi_j)_{L^2}| < \infty$ and $\sum_j |(\widehat{A}_2\phi_j|\chi_j)_{L^2}| < \infty$, then we have, using the triangle inequal-

https://doi.org/10.1515/9783110722772-012

ity,

$$\sum_j |((\widehat{A}_1 + \widehat{A}_2)\phi_j|\chi_j)_{L^2}| \le \sum_j |(\widehat{A}_1\phi_j|\chi_j)_{L^2}| + \sum_j |(\widehat{A}_2\phi_j|\chi_j)_{L^2}| < \infty$$

so that $\widehat{A}_1 + \widehat{A}_2 \in \mathcal{L}_1(L^2(\mathbb{R}^n))$; that $\lambda\widehat{A} \in \mathcal{L}_1(L^2(\mathbb{R}^n))$ if $\widehat{A} \in \mathcal{L}_1(L^2(\mathbb{R}^n))$ and $\lambda \in \mathbb{C}$ is clear.

12.1.2 Proof of the equivalence of the definitions

Let us show that the two definitions 144 and 146 of trace class operators are equivalent.

We will need the following useful result about absolute values of trace class operators:

Proposition 147. *The operator $\widehat{A}$ is of trace class if and only if its absolute value $|\widehat{A}| = (\widehat{A}^*\widehat{A})^{1/2}$ is of trace class.*

Proof. In view of the polar decomposition theorem, there exists a unitary operator U on $L^2(\mathbb{R}^n)$ such that $\widehat{A} = \widehat{U}(\widehat{A}^*\widehat{A})^{1/2}$. Supposing $\widehat{A}$ is of trace class, then

$$\sum_{j,k} (\widehat{U}(\widehat{A}^*\widehat{A})^{1/2}\psi_j|\phi_k)_{L^2} < \infty$$

for some orthonormal bases $(\psi_j)_j$ and $(\phi_k)_k$, that is

$$\sum_{j,k} ((\widehat{A}^*\widehat{A})^{1/2}\psi_j|\phi_k')_{L^2} < \infty$$

where $\phi_k' = \widehat{U}^{-1}\phi_k$. Since $\widehat{U}$ is unitary, $(\phi_k')_k$ is also an orthonormal basis, hence $|\widehat{A}| = (\widehat{A}^*\widehat{A})^{1/2}$ is of trace class. If, conversely, $|\widehat{A}|$ is of trace class, then so is $\widehat{A} = \widehat{U}(\widehat{A}^*\widehat{A})^{1/2}$, replacing again $(\phi_k)_k$ with $(\widehat{U}^{-1}\phi_k')_k$. □

Proposition 148. *Definitions* 144 *and* 146 *define the same class of operators.*

Proof. In what follows, (ψ_j) and $(\phi_j)_j$ are two arbitrary orthonormal bases of $L^2(\mathbb{R}^n)$. Let us first show that

$$\text{Definition 144} \Longrightarrow \text{Definition 146.}$$

Suppose that $\widehat{A} = \widehat{C}^*\widehat{B}$ where $\widehat{B}$ and $\widehat{C}$ are both Hilbert–Schmidt operators. Since $(\widehat{A}\psi_j|\phi_k)_{L^2} = (\widehat{B}\psi_j|\widehat{C}\phi_k)_{L^2}$, we have, using successively the triangle and the Cauchy–Schwarz inequalities, together with the trivial inequality $bc \le \frac{1}{2}(b^2 + c^2)$,

$$\begin{aligned}\left|\sum_{j,k} (\widehat{A}\psi_j|\phi_k)_{L^2}\right| &\le \sum_{j,k} |(\widehat{B}\psi_j|\widehat{C}\phi_k)_{L^2}| \\ &\le \sum_{j,k} \|\widehat{B}\psi_j\|_{L^2}\|\widehat{C}\phi_k\|_{L^2}\end{aligned}$$

$$\leq \frac{1}{2}\Big(\sum_j \|\widehat{B}\psi_j\|_{L^2}^2 + \sum_k \|\widehat{C}\phi_k\|_{L^2}^2\Big)$$
$$= \frac{1}{2}(\|\widehat{B}\|_2^2 + \|\widehat{C}\|_2^2) < \infty,$$

the last equality since $\widehat{B}$ and $\widehat{C}$ are both Hilbert–Schmidt operators. We have thus proven that $\widehat{A}$ satisfies the condition in Definition 144. Let us next show that, conversely,

$$\text{Definition 146} \Longrightarrow \text{Definition 144}.$$

We choose orthonormal bases (ψ_j) and $(\phi_j)_j$ such that

$$\sum_{j,k} |(|\widehat{A}|\psi_j|\phi_k)_{L^2}| < \infty \tag{12.2}$$

(recall from Proposition 147 that $\widehat{A}$ is of trace class if and only if $\widehat{A}|$ is). Using the polar decomposition theorem, we can write $\widehat{A} = \widehat{U}(\widehat{A}^*\widehat{A})^{1/2} = \widehat{U}|\widehat{A}|$ where U is unitary on $L^2(\mathbb{R}^n)$. Setting $\widehat{C}^* = U(\widehat{A}^*\widehat{A})^{1/4}$ and $\widehat{B} = (\widehat{A}^*\widehat{A})^{1/4}$, we have $\widehat{A} = \widehat{C}^*\widehat{B}$; let us show that $\widehat{C}$ and $\widehat{B}$ are both Hilbert–Schmidt operators; the proof will follow since $\widehat{A}$ will then satisfy the condition in Definition 144. To show that $\widehat{C} \in \mathcal{L}_2(L^2(\mathbb{R}^n))$, it suffices to prove that $\widehat{C}^* \in \mathcal{L}_2(L^2(\mathbb{R}^n))$, that is

$$\sum_j |(\widehat{C}^*\psi_j|\widehat{C}^*\phi_j)_{L^2}| < \infty.$$

We have, since the absolute values of Hilbert–Schmidt operators also are Hilbert–Schmidt (Proposition 140),

$$\sum_{j,k} |(\widehat{C}^*\psi_j|\widehat{C}^*\phi_k)_{L^2}| = \sum_{j,k} |(\widehat{C}\widehat{C}^*\psi_j|\phi_k)_{L^2}|$$
$$= \sum_{j,k} |((\widehat{A}^*\widehat{A})^{1/2}\psi_j|\phi_k)_{L^2}|$$
$$= \sum_{j,k} |(|\widehat{A}|\psi_j|\phi_k)_{L^2}| < \infty,$$

hence $\widehat{C} \in \mathcal{L}_2(L^2(\mathbb{R}^n))$ as claimed. To show that we also have $\widehat{B} \in \mathcal{L}_2(L^2(\mathbb{R}^n))$, the argument is exactly the same, noting that $\widehat{B}^*\widehat{B} = \widehat{C}^*\widehat{C} = |\widehat{A}|$. □

Let us next show that in Definition 146 the choice of orthonormal bases in (12.1) has no importance and that it suffices to prove that, if condition (12.1) holds for one particular choice of orthonormal bases, then it holds for all. This will in addition allow us to define unambiguously the trace of a trace class operator.

Proposition 149. (i) *An operator* $\widehat{A} \in \mathcal{B}(L^2(\mathbb{R}^n))$ *is of trace class if and only if*

$$\sum_{j,k} |(\widehat{A}\psi_j|\phi_k)_{L^2}| < \infty \tag{12.3}$$

for every pair $((\psi_j)_j, (\phi_j)_j)$ *of bases of* $L^2(\mathbb{R}^n)$. (ii) *When this is the case, we have*

$$\sum_j (\widehat{A}\psi_j|\psi_j)_{L^2} = \sum_j (\widehat{A}\phi_j|\phi_j)_{L^2} < \infty \tag{12.4}$$

for all orthonormal bases $(\psi_j)_j$ *and* $(\phi_j)_j$ *of* $L^2(\mathbb{R}^n)$ *and where both series are absolutely convergent.*

Proof. (i) Let $((\psi_j)_j, (\phi_j)_j)$ and $((\psi_j')_j, (\phi_j')_j)$ be two pairs of orthonormal bases of $L^2(\mathbb{R}^n)$. Writing Fourier expansions

$$\psi_i' = \sum_j (\psi_i'|\psi_j)_{L^2}\psi_j, \quad \phi_m' = \sum_k (\phi_m'|\phi_k)_{L^2}\phi_k,$$

we have

$$(\widehat{A}\psi_i'|\phi_m')_{L^2} = \sum_{j,k} (\psi_i'|\psi_j)_{L^2}\overline{(\phi_m'|\phi_k)_{L^2}}(\widehat{A}\psi_j|\phi_k)_{L^2}, \tag{12.5}$$

and hence

$$|(\widehat{A}\psi_i'|\phi_m')_{L^2}| \leq \sum_{j,k} |(\psi_i'|\psi_j)_{L^2}|\,|(\phi_m'|\phi_k)_{L^2}|\,|(\widehat{A}\psi_j|\phi_k)_{L^2}|.$$

Summing with respect to the indices i, m, this inequality yields

$$\sum_{i,m} |(\widehat{A}\psi_i'|\phi_m')_{L^2}| \leq \sum_{j,k}\Big(\sum_{i,m} |(\psi_i'|\psi_j)_{L^2}|\,|(\phi_m'|\phi_k)_{L^2}|\Big)|(\widehat{A}\psi_j|\phi_k)_{L^2}|$$

(the interchange of summations is justified since the series consist of positive terms), and hence

$$\sum_{i,m} \|(\widehat{A}\psi_i'|\phi_m')_{L^2}\| \leq \frac{1}{2}\sum_{j,k}\Big(\sum_i |(\psi_i'|\psi_j)_{L^2}^2| + \sum_m |(\phi_m'|\phi_k)_{L^2}^2|\Big)|(\widehat{A}\psi_j|\phi_k)_{L^2}|.$$

Now,

$$\sum_i |(\psi_i'|\psi_j)_{L^2}^2| = \|\psi_j\|_{L^2}^2 = 1, \quad \sum_m |(\phi_m'|\phi_k)_{L^2}^2| = \|\phi_k\|_{L^2}^2 = 1,$$

and hence the inequality becomes

$$\sum_{i,m} \|(\widehat{A}\psi_i'|\phi_m')_{L^2}\| \leq \sum_{j,k} |(\widehat{A}\psi_j|\phi_k)_{L^2}|.$$

Hence condition (12.1) holds for the bases $((\psi'_j)_j, (\phi'_j)_j)$ if it holds for $((\psi_j)_j, (\phi_j)_j)$. (ii) In view of the first part of the proposition, if $\widehat{A}$ is of trace class, then

$$\left|\sum_j (\widehat{A}\psi_j|\psi_j)_{L^2}\right| \leq \sum_j |(\widehat{A}\psi_j|\psi_j)_{L^2}| \leq \sum_{j,k} |(\widehat{A}\psi_j|\psi_k)_{L^2}| < \infty,$$

so the series $\sum_j (\widehat{A}\psi_j|\psi_j)_{L^2}$ is absolutely convergent. The equality (12.5) implies, in particular, that

$$(\widehat{A}\phi_i|\phi_i)_{L^2} = \sum_{j,k} (\phi_i|\psi_j)_{L^2}\overline{(\phi_i|\psi_k)_{L^2}}(\widehat{A}\psi_j|\psi_k)_{L^2},$$

and hence

$$\sum_i (\widehat{A}\phi_i|\phi_i)_{L^2} = \sum_{j,k}\left(\sum_i (\phi_i|\psi_j)_{L^2}\overline{(\phi_i|\psi_k)_{L^2}}\right)(\widehat{A}\psi_j|\psi_k)_{L^2}.$$

Since we have

$$\sum_i (\phi_i|\psi_j)_{L^2}\overline{(\phi_i|\psi_k)_{L^2}} = (\psi_j|\psi_k) = \delta_{jk},$$

the equality becomes

$$\sum_i (\widehat{A}\phi_i|\phi_i)_{L^2} = \sum_j (\widehat{A}\psi_j|\psi_j)_{L^2},$$

which was to be proven. □

12.1.3 Definition of the trace

In many texts, one finds the following definition: Let $\widehat{A}$ be a positive semidefinite operator on some infinitely dimensional separable Hilbert space H, and let $(\psi_j)_j$ be an orthonormal basis of H. Then

$$\operatorname{Tr}(\widehat{A}) = \sum_j (\widehat{A}\psi_j|\psi_j)_{L^2} \leq \infty$$

is called the trace of $\widehat{A}$, and $\operatorname{Tr}(\widehat{A})$ does not depend on the choice of $(\psi_j)_j$. Doing so, one has to do thereafter some juggling to define the trace of an arbitrary operator (when it exists). We prefer the following approach:

Definition 150. Let $\widehat{A} \in \mathcal{L}_1(L^2(\mathbb{R}^n))$ be a trace class operator. By definition, the trace of $\widehat{A}$ is the real number

$$\operatorname{Tr}(\widehat{A}) = \sum_j (\widehat{A}\psi_j|\psi_j)_{L^2} \tag{12.6}$$

whose value is independent of the choice of orthonormal basis $(\psi_j)_j$ of $L^2(\mathbb{R}^n)$.

It immediately follows, from this definition, that the trace satisfies the two following properties: If $\widehat{A} \in \mathcal{L}_1(L^2(\mathbb{R}^n))$, then

$$\widehat{A} \geq 0 \Longrightarrow \mathrm{Tr}(\widehat{A}) \geq 0; \quad \mathrm{Tr}(\widehat{A}^*) = \overline{\mathrm{Tr}(\widehat{A})} \tag{12.7}$$

(in particular $\mathrm{Tr}(\widehat{A})$ is a real number if $\widehat{A}^* = \widehat{A}$). Also note that, following the spectral theorem for compact operators, we have

$$\mathrm{Tr}(\widehat{A}) = \sum_j \lambda_j \tag{12.8}$$

where the λ_j are the eigenvalues of $\widehat{A}$.

12.2 Properties of the trace class

12.2.1 Trace class operators form an ideal

Let us first show that the trace class is invariant under conjugation with unitary operators:

Proposition 151. *Let $\widehat{A} \in \mathcal{L}_1(L^2(\mathbb{R}^n))$ and $\widehat{U}$ be an unitary operator on $L^2(\mathbb{R}^n)$. Then, $\widehat{U}^*\widehat{A}\widehat{U} \in \mathcal{L}_1(L^2(\mathbb{R}^n))$ and has the same trace as $\widehat{A}$:*

$$\mathrm{Tr}(\widehat{U}^*\widehat{A}\widehat{U}) = \mathrm{Tr}(\widehat{A}). \tag{12.9}$$

Proof. The operator $\widehat{A}$ is of trace class if and only if $\sum_j (\widehat{A}\psi_j|\psi_j)_{L^2} < \infty$ for one (and, hence, every) orthonormal basis $(\psi_j)_j$ of $L^2(\mathbb{R}^n)$. Since $(\widehat{U}^*\widehat{A}\widehat{U}\psi_j|\psi_j)_{L^2} = (\widehat{A}\widehat{U}\psi_j|\widehat{U}\psi_j)_{L^2}$ and $(\widehat{U}\psi_j)_j$ also is an orthonormal basis of L^2, it follows that $\widehat{U}^*\widehat{A}\widehat{U}$ is of trace class; formula (12.9) now follows: We have

$$\begin{aligned} \mathrm{Tr}(\widehat{U}^*\widehat{A}\widehat{U}) &= \sum_j (\widehat{U}^*\widehat{A}\widehat{U}\psi_j|\psi_j)_{L^2} \\ &= \sum_j (\widehat{A}\widehat{U}\psi_j|\widehat{U}\psi_j)_{L^2} = \mathrm{Tr}(\widehat{A}). \end{aligned}$$

□

Note that formula (12.9) reflects the freedom we have in choosing the orthonormal basis in which the trace is calculated (formula (12.6)).

Trace class operators form an ideal in the bounded operators:

Proposition 152. *The set $\mathcal{L}_1(L^2(\mathbb{R}^n))$ of all trace class operators on $L^2(\mathbb{R}^n)$ is a two-sided ideal in $\mathcal{B}(L^2(\mathbb{R}^n))$: If $\widehat{A} \in \mathcal{L}_1(L^2(\mathbb{R}^n))$ and $\widehat{B} \in \mathcal{B}(L^2(\mathbb{R}^n))$, then $\widehat{A}\widehat{B} \in \mathcal{L}_1(L^2(\mathbb{R}^n))$ and $\widehat{B}\widehat{A} \in \mathcal{L}_1(L^2(\mathbb{R}^n))$, and we have*

$$\mathrm{Tr}(\widehat{A}\widehat{B}) = \mathrm{Tr}(\widehat{B}\widehat{A}). \tag{12.10}$$

Proof. Let us show that $\widehat{B}\widehat{A} \in \mathcal{L}_1(L^2(\mathbb{R}^n))$ if $\widehat{A} \in \mathcal{L}_1(L^2(\mathbb{R}^n))$ and $\widehat{B}$ is a bounded operator on $L^2(\mathbb{R}^n)$. Recall that the boundedness of $\widehat{B}$ is equivalent to the existence of a number $C \geq 0$ such that $\|\widehat{B}\psi\| \leq C\|\psi\|$ for all $\psi \in L^2(\mathbb{R}^n)$. Let now $(\psi_j)_j$ and $(\phi_j)_j$ be two orthonormal bases of $L^2(\mathbb{R}^n)$; writing $(\widehat{B}\widehat{A}\psi_j|\phi_j)_{L^2} = (\widehat{A}\psi_j|\widehat{B}^*\phi_j)_{L^2}$ and applying Bessel's equality to $(\widehat{A}\psi_j|\widehat{B}^*\phi_j)_{L^2}$, we get

$$(\widehat{B}\widehat{A}\psi_j|\phi_j)_{L^2} = \sum_k (\widehat{A}\psi_j|\phi_k)_{L^2}\overline{(\widehat{B}^*\phi_j|\psi_k)_{L^2}}. \tag{12.11}$$

Using the Cauchy–Schwarz inequality, we have, since $\|\widehat{A}^*\phi_j\| \leq C$,

$$\overline{|(\widehat{B}^*\phi_j|\psi_k)_{L^2}} \leq \|\widehat{B}^*\phi_j\|_{L^2}\,\|\psi_k\|_{L^2} \leq C,$$

and hence

$$\begin{aligned}|(\widehat{B}\widehat{A}\psi_j|\phi_j)_{L^2}| &\leq \sum_k |(\widehat{A}\psi_j|\phi_k)_{L^2}(\widehat{B}^*\phi_j|\psi_k)^*_{L^2}| \\ &\leq C\sum_k |(\widehat{A}\psi_j|\phi_k)_{L^2}|.\end{aligned}$$

Summing this inequality with respect to the index j yields, since $\widehat{A}$ is of trace class,

$$\left|\sum_j (\widehat{B}\widehat{A}\psi_j|\phi_j)_{L^2}\right| \leq C\sum_{j,k} |(\widehat{A}\psi_j|\phi_k)_{L^2}| < \infty,$$

hence $\widehat{B}\widehat{A}$ is of trace class as claimed. That $\widehat{A}\widehat{B}$ also is of trace class is immediate, noting that we can write $\widehat{A}\widehat{B} = (\widehat{B}^*\widehat{A}^*)^*$ and using the fact that the trace class is preserved by passage to the adjoint. There remains to prove the trace equality (12.10). Choosing $(\psi_j)_j = (\phi_j)_j$, the Bessel equality (12.11) yields

$$(\widehat{B}\widehat{A}\psi_j|\psi_j)_{L^2} = \sum_k (\widehat{A}\psi_j|\psi_k)_{L^2}\overline{(\widehat{B}^*\psi_j|\psi_k)_{L^2}},$$

and hence, summing over j,

$$\operatorname{Tr}(\widehat{A}\widehat{B}) = \sum_j (\widehat{A}\widehat{B}\psi_j|\psi_j)_{L^2} = \sum_{j,k} (\widehat{A}\psi_j|\psi_k)_{L^2}\overline{(\widehat{B}^*\psi_j|\psi_k)_{L^2}}. \tag{12.12}$$

Similarly,

$$\begin{aligned}(\widehat{A}\widehat{B}\psi_j|\psi_j)_{L^2} &= \sum_k (\widehat{B}\psi_j|\psi_k)_{L^2}\overline{(\widehat{A}^*\psi_j|\psi_k)_{L^2}} \\ &= \sum_k (\widehat{A}\psi_k|\psi_j)_{L^2}\overline{(\widehat{B}^*\psi_k|\psi_j)_{L^2}},\end{aligned}$$

hence

$$\operatorname{Tr}(\widehat{B}\widehat{A}) = \sum_j (\widehat{A}\widehat{B}\psi_j|\psi_j)_{L^2} = \sum_{j,k} (\widehat{A}\psi_k|\psi_j)_{L^2}\overline{(\widehat{B}^*\psi_k|\psi_j)_{L^2}}. \tag{12.13}$$

Comparing (12.12) and (12.13), we have $\operatorname{Tr}(\widehat{A}\widehat{B}) = \operatorname{Tr}(\widehat{B}\widehat{A})$ as we set out to prove. □

12.2.2 The trace class norm

Let us define the trace class norm:

Definition 153. The trace class norm is the mapping $\|\cdot\|_1 : \mathcal{L}_1(L^2(\mathbb{R}^n)) \longrightarrow \mathbb{R}_+$ defined by

$$\|\widehat{A}\|_1 = \operatorname{Tr}(|\widehat{A}|)$$

where $|\widehat{A}| = (\widehat{A}^*\widehat{A})^{1/2}$ is the modulus of $\widehat{A}$.

Clearly, $\|\widehat{A}\|_1 \geq 0$ and $\|\alpha\widehat{A}\|_1 = |\alpha|\,\|\widehat{A}\|_1$ for $\alpha \in \mathbb{C}$. Also, $\|\widehat{A}\|_1 = 0$ implies that $|\widehat{A}| = 0$, hence $\widehat{A} = 0$. There remains to prove the triangle inequality

$$\|\widehat{A} + \widehat{B}\|_1 \leq \|\widehat{A}\|_1 + \|\widehat{B}\|_1. \tag{12.14}$$

For this, we will need the formula

$$\|\widehat{A}\|_1 = \sup\{\operatorname{Tr}(\widehat{X}\widehat{A}) : \widehat{X} \in \mathcal{L}_1(L^2(\mathbb{R}^n)), \|\widehat{X}\| \leq 1\}, \tag{12.15}$$

which follows from the double inequality

$$|\operatorname{Tr}(\widehat{X}\widehat{A})| \leq \|\widehat{X}\widehat{A}\|_1 \leq \|\widehat{X}\|\,\|\widehat{A}\|_1$$

(we leave the proof of (12.15) to the reader as a pleasant exercise). Using (12.15), we get

$$\begin{aligned}
\|\widehat{A} + \widehat{B}\|_1 &= \sup_{\|\widehat{X}\|\leq 1} |\operatorname{Tr}(\widehat{X}\widehat{A}) + \operatorname{Tr}(\widehat{X}\widehat{B})| \\
&\leq \sup_{\|\widehat{X}\|\leq 1} (|\operatorname{Tr}(\widehat{X}\widehat{A})| + |\operatorname{Tr}(\widehat{X}\widehat{B})|) \\
&\leq \sup_{\|\widehat{Y}\|\leq 1, \|\widehat{Z}\|\leq 1} (|\operatorname{Tr}(\widehat{Y}\widehat{A})| + |\operatorname{Tr}(\widehat{Z}\widehat{B})|).
\end{aligned}$$

We can rewrite the last inequality as

$$\begin{aligned}
\|\widehat{A} + \widehat{B}\|_1 &\leq \sup_{\|\widehat{Y}\|\leq 1} (|\operatorname{Tr}(\widehat{Y}\widehat{A})|) + \sup_{\|\widehat{Z}\|\leq 1} (|\operatorname{Tr}(\widehat{Z}\widehat{B})|) \\
&= \|\widehat{A}\|_1 + \|\widehat{B}\|_1,
\end{aligned}$$

which ends the proof of the triangle inequality for the trace norm.

Proposition 154. *$\mathcal{L}_1(L^2(\mathbb{R}^n))$ is a Banach algebra for the trace class norm.*

We omit the proof of this property here; see the comments and references section.

12.3 Trace formulas

12.3.1 A product formula

Recall that every trace class operator is the product of two Hilbert–Schmidt operators.

Proposition 155. *Let $\widehat{A} = \mathrm{Op_W}(a)$ and $\widehat{B} = \mathrm{Op_W}(b)$ be Hilbert–Schmidt operators. The trace of $\widehat{A}\widehat{B} \in \mathcal{L}_1(L^2(\mathbb{R}^n))$ is given by*

$$\operatorname{Tr}(\widehat{A}\widehat{B}) = \left(\frac{1}{2\pi\hbar}\right)^n \int_{\mathbb{R}^{2n}} a(z)b(z)dz. \tag{12.16}$$

Proof. Since $\widehat{A}$ and $\widehat{B}$ are Hilbert–Schmidt operators, we have $a \in L^2(\mathbb{R}^{2n})$ and $b \in L^2(\mathbb{R}^{2n})$. Let $(\psi_j)_j$ be an orthonormal basis of $L^2(\mathbb{R}^n)$; we have

$$\operatorname{Tr}(\widehat{A}\widehat{B}) = \sum_{j=1}^{\infty}(\widehat{A}\widehat{B}\psi_j|\psi_j)_{L^2} = \sum_{j=1}^{\infty}(\widehat{B}\psi_j|\widehat{A}^*\psi_j)_{L^2}.$$

Expanding $\widehat{B}\psi_j$ and $\widehat{A}^*\psi_j$ in the basis $(\psi_j)_j$, we have

$$\widehat{B}\psi_j = \sum_{k=1}^{\infty}(\widehat{B}\psi_j|\psi_k)_{L^2}\psi_k$$

$$\widehat{A}^*\psi_j = \sum_{\ell=1}^{\infty}(\psi_j|\widehat{A}\psi_\ell)_{L^2}\psi_\ell,$$

and hence

$$(\widehat{B}\psi_j|\widehat{A}^*\psi_j)_{L^2} = \sum_{k=1}^{\infty}(\widehat{B}\psi_j|\psi_k)_{L^2}\overline{(\widehat{A}\psi_j|\psi_k)_{L^2}}.$$

Observing that we have

$$(\widehat{A}\psi_j|\psi_k)_{L^2} = \int_{\mathbb{R}^{2n}} a(z)W(\psi_j,\psi_k)(z)dz = (a|W(\psi_k,\psi_j))_{L^2(\mathbb{R}^{2n})}$$

$$(\widehat{B}\psi_j|\psi_k)_{L^2} = \int_{\mathbb{R}^{2n}} b(z)W(\psi_j,\psi_k)(z)dz = (b|W(\psi_k,\psi_j))_{L^2(\mathbb{R}^{2n})},$$

the previous above can be rewritten as

$$(\widehat{B}\psi_j|\widehat{A}^*\psi_j)_{L^2} = \sum_{k=1}^{\infty}(a|W(\psi_k,\psi_j))_{L^2(\mathbb{R}^{2n})}(b|W(\psi_k,\psi_j))_{L^2(\mathbb{R}^{2n})}.$$

Recalling (Proposition 64) that, if $(\psi_j)_j$ is an orthonormal basis of $L^2(\mathbb{R}^n)$, then the vectors $\Phi_{j,k} = (2\pi\hbar)^{n/2}W(\psi_k,\psi_j)$ form an orthonormal basis of $L^2(\mathbb{R}^{2n})$, we have

$$\operatorname{Tr}(\widehat{A}\widehat{B}) = \sum_j(\widehat{B}\psi_j|\widehat{A}^*\psi_j)_{L^2(\mathbb{R}^n)}$$

$$= \left(\frac{1}{2\pi\hbar}\right)^n \sum_{j,k} (a|\Phi_{j,k})_{L^2(\mathbb{R}^{2n})}(b|\Phi_{j,k})_{L^2(\mathbb{R}^{2n})}$$

$$= \left(\frac{1}{2\pi\hbar}\right)^n (a|\overline{b})_{L^2(\mathbb{R}^{2n})},$$

and this proves formula (12.16). □

Remark 156. Let $\widehat{A}$ and $\widehat{B}$ be two Hilbert–Schmidt operators on $L^2(\mathbb{R}^n)$; the formula

$$(\widehat{A}|\widehat{B})_{\mathcal{L}_2} = \operatorname{Tr}(\widehat{A}^*\widehat{B})$$

defines a scalar product on $\mathcal{L}_2(L^2(\mathbb{R}^n))$.

12.3.2 Relationship between the trace and the Weyl symbol

Let $\widehat{A}$ be a self-adjoint trace class operator on $L^2(\mathbb{R}^n)$. In view of the spectral theorem for compact operators, there exists a most countable set (ψ_j) of orthonormal eigenvectors of $\widehat{A}$ with eigenvalues λ_j such that, if ψ_j is an eigenvector for λ_j, then

$$\widehat{A}\psi = \sum_j \lambda_j (\psi|\psi_j)_{L^2}\psi_j \tag{12.17}$$

where $(\lambda_j)_j \in \ell^1(\mathbb{N})$. As a consequence, the Weyl symbol of $\widehat{A}$ is given by the absolutely convergent series

$$a(z) = (2\pi\hbar)^n \sum_j \lambda_j W\psi_j(z). \tag{12.18}$$

We omit the proof of the following result due to Shubin:

Proposition 157. *Let a be in the Shubin symbol class $\Gamma_\rho^m(\mathbb{R}^{2n})$ with $m < -2n$. Then the Weyl operator $\widehat{A} = \operatorname{Op_W}(a)$ is of trace class and has trace*

$$\operatorname{Tr}(\widehat{A}) = \left(\frac{1}{2\pi\hbar}\right)^n \int_{\mathbb{R}^{2n}} a(z)dz. \tag{12.19}$$

If one assumes from the beginning that $\operatorname{Op_W}(a)$ is of trace class, one has the following stronger result:

Proposition 158. *Let $\widehat{A} = \operatorname{Op_W}(a)$ be a trace class operator. If $a \in L^1(\mathbb{R}^{2n})$, then*

$$\operatorname{Tr}(\widehat{A}) = \left(\frac{1}{2\pi\hbar}\right)^n \int_{\mathbb{R}^{2n}} a(z)dz. \tag{12.20}$$

Proof. We notice that the trace formula (12.20) can be rewritten as

$$\operatorname{Tr}(\widehat{A}) = a_\sigma(0)$$

where $a_\sigma = F_\sigma a$ is the symplectic Fourier transform of $a \in L^1(\mathbb{R}^{2n})$. Let us write $\widehat{A} = \widehat{B}\widehat{C}$ where $\widehat{B}$ and $\widehat{C}$ are Hilbert–Schmidt operators. Using formula (12.16) in Proposition 155, we have

$$\operatorname{Tr}(\widehat{A}) = \left(\frac{1}{2\pi\hbar}\right)^n \int_{\mathbb{R}^{2n}} b(z)c(z)dz.$$

Let us show that

$$a_\sigma(0) = \left(\frac{1}{2\pi\hbar}\right)^n \int_{\mathbb{R}^{2n}} b(z)c(z)dz;$$

formula (12.20) will follow. We have, in view of the composition formula (5.16),

$$a_\sigma(z) = \left(\frac{1}{2\pi\hbar}\right)^n \int_{\mathbb{R}^{2n}} e^{\frac{i}{2\hbar}\sigma(z,z')} b_\sigma(z-z')c_\sigma(z')dz',$$

and hence

$$\begin{aligned} a_\sigma(0) &= \left(\frac{1}{2\pi\hbar}\right)^n \int_{\mathbb{R}^{2n}} (b_\sigma)^\vee(z)c_\sigma(z)dz \\ &= \left(\frac{1}{2\pi\hbar}\right)^n ((b_\sigma)^\vee|\overline{c_\sigma})_{L^2(\mathbb{R}^{2n})} \end{aligned}$$

with $(b_\sigma)^\vee(z) = b_\sigma(-z)$. Noting that $(b_\sigma)^\vee = (b^\vee)_\sigma$ and $\overline{c_\sigma} = (\overline{c^\vee})_\sigma$, we thus have, since the symplectic Fourier transform is unitary,

$$\begin{aligned} a_\sigma(0) &= \left(\frac{1}{2\pi\hbar}\right)^n ((b^\vee)_\sigma|(\overline{c^\vee})_\sigma)_{L^2(\mathbb{R}^{2n})} \\ &= \left(\frac{1}{2\pi\hbar}\right)^n (b^\vee|\overline{c^\vee})_{L^2(\mathbb{R}^{2n})} \\ &= \left(\frac{1}{2\pi\hbar}\right)^n \int_{\mathbb{R}^{2n}} b(z)c(z)dz, \end{aligned}$$

which was to be proven. □

Remark 159. One can show that then that, if $a \in L^1(\mathbb{R}^n)$ but $a \notin L^2(\mathbb{R}^n)$, $\widehat{A}$ is not a trace class operator.

12.4 Comments and references

There are many equivalent ways to address the topic of trace class operators, whose importance will become clear in next chapter where we explain the density operator formalism. We have been remotely following the approach in [31], with corrections, modifications, and complements. The elementary theory of Hilbert–Schmidt and trace class operators as studied here is not difficult (it is essentially linear algebra in infinite-dimensional vector spaces), but it involves various tricks and juggling with cascades of indices. An elegant (but very concise) description apparently avoiding these calculations can be found in Hörmander [46], § 19.1, or in Shubin [66] (Appendix 3). Again, MacCluer [57] and Blanchard and Brüning [6] are recent good introductory texts.

For the proof of Proposition 157, see Shubin [66], § 27.1 (Proposition 27.2(2)). The proof of this proposition is due to Du and Wong [23] (Theorem 2.4).

The proof of the triangle inequality (12.14) and of formula (12.15) for the trace norm is due to Shubin [66], § A.3.4, in the form presented here; one also finds in this reference a complete proof of the completeness of the trace class.

13 The quantum Bochner theorem

The quantum Bochner theorem deals with positivity questions for trace class operators. It is a difficult topic, which is still being investigated in the research literature.

13.1 The positivity question

Recall that every continuous linear operator $\widehat{A} : \mathcal{S}(\mathbb{R}^n) \longrightarrow \mathcal{S}'(\mathbb{R}^n)$ can be expressed in terms of the symplectic Fourier transform of its Weyl symbol and the Heisenberg displacement operator as

$$\widehat{A}\psi = \left(\frac{1}{2\pi\hbar}\right)^n \int_{\mathbb{R}^{2n}} a_\sigma(z)\widehat{D}(z)\psi dz \tag{13.1}$$

for $\psi \in \mathcal{S}(\mathbb{R}^n)$ and, equivalently, in terms of the reflection operator

$$\widehat{A}\psi = \left(\frac{1}{\pi\hbar}\right)^n \int_{\mathbb{R}^{2n}} a(z)\widehat{R}(z)\psi dz. \tag{13.2}$$

In this chapter, we investigate which conditions a or a_σ should satisfy for a *trace class operator* $\widehat{A}$ to be semidefinite positive, i. e.,

$$(\widehat{A}\psi|\psi) \geq 0 \quad \text{for all } \psi \in L^2(\mathbb{R}^n).$$

This question is actually much more subtle than it seems, and the main culprit is the difficulty one has to give tractable conditions on the symbol a; much literature has been devoted to the subject (see the comments and references section at the end the chapter).

13.2 Bochner's theorem

A famous theorem of Bochner says that a (complex valued) function f on $\mathbb{R}^n$, continuous at the origin and such that $f(0) = 1$, is the Fourier transform of a probability density on $\mathbb{R}^n$ if and only if it is of *positive type*, that is, if for all choices of points $z_1, \dots, z_N \in \mathbb{R}^n$, the $N \times N$ matrix

$$F_{(N)} = (f(z_j - z_k))_{1\leq j,k\leq N} \tag{13.3}$$

is positive semidefinite (that is, the eigenvalues of $F_{(N)}$ are all ≥ 0). Let us introduce a modification of the symplectic Fourier transform F_σ.

https://doi.org/10.1515/9783110722772-013

Definition 160. The *reduced symplectic Fourier transform* $F_\diamond$ is defined, for $a \in L^1(\mathbb{R}^{2n})$, by

$$a_\diamond(z) = F_\diamond a(z) = \int e^{-i\sigma(z,z')} a(z') d^{2n} z'. \tag{13.4}$$

The functions $a_\diamond$ and $a_\sigma = F_\sigma a$ are related by the formula

$$a_\diamond(z) = (2\pi\hbar)^n a_\sigma(\hbar z). \tag{13.5}$$

With this notation, Bochner's theorem on Fourier transforms of probability measures can be restated in the following way: A real function ρ on $\mathbb{R}^{2n}$ is a probability density if and only if its reduced symplectic Fourier transform $\rho_\diamond$ is continuous, $\rho_\diamond(0) = 1$, and, for all choices of $z_1, \dots, z_N \in \mathbb{R}^{2n}$, the $N \times N$ matrix Λ whose entries are the complex numbers $\rho_\diamond(z_j - z_k)$ is positive semidefinite:

$$\Lambda = (\rho_\diamond(z_j - z_k))_{1\leq j,k\leq N} \geq 0. \tag{13.6}$$

When condition (13.6) is satisfied, one says that the reduced symplectic Fourier transform $\rho_\diamond$ is of *positive type*.

13.3 The quantum case

13.3.1 The KLM conditions

To deal with the quantum case, we need the following definition:

Definition 161 (KLM conditions). Let $a \in L^1(\mathbb{R}^{2n})$; we say that $a_\diamond$ is of *ħ-positive type if for every integer N the* $N \times N$ matrix $\Lambda_{(N)}$ with entries

$$\Lambda_{jk} = e^{-\frac{i\hbar}{2}\sigma(z_j,z_k)} a_\diamond(z_j - z_k)$$

is positive semidefinite for all choices of $(z_1, z_2, \dots, z_N) \in (\mathbb{R}^{2n})^N$:

$$\Lambda_{(N)} = (\Lambda_{jk})_{1\leq j,k\leq N} \geq 0. \tag{13.7}$$

The condition (13.7) is equivalent to the polynomial inequalities

$$\sum_{1\leq j,k\leq N} \zeta_j \overline{\zeta_k} e^{-\frac{i\hbar}{2}\sigma(z_j,z_k)} a_\diamond(z_j - z_k) \geq 0 \tag{13.8}$$

for all $N \in \mathbb{N}$, $\zeta_j, \zeta_k \in \mathbb{C}$, and $z_j, z_k \in \mathbb{R}^{2n}$.

It is easy to see that this implies $a_\diamond(-z) = \overline{a_\diamond(z)}$, and therefore a must be real-valued.

Remark 162. The conditions (13.7) are often called the "KLM conditions" in the literature following the seminal work of Kastler, Loupias, and Miracle-Sole (see the comments and references at the end of this chapter).

Remark 163. One can prove that, if the function $a : \mathbb{R}^{2n} \longrightarrow \mathbb{C}$ is of *ℏ-positive type, then* $a \in L^2(\mathbb{R}^{2n})$.

13.3.2 The main result; statement and proof

We will need the following lemma, due to Schur. Recall that the Hadamard (or Schur) product of two matrices $M'_{(N)} = (M'_{jk})_{1\leq j,k\leq N}$ and $M''_{(N)} = (M''_{jk})_{1\leq j,k\leq N}$ is defined by entrywise multiplication:

$$M'_{(N)} \circ M''_{(N)} = (M'_{jk}M''_{jk})_{1\leq j,k\leq N}.$$

We will need the following classical lemma:

Lemma 164. *Let* $M_{(N)} = (M_{jk})_{1\leq j,k\leq N}$ *be the Hadamard product* $M'_{(N)} \circ M''_{(N)}$ *of the matrices* $M'_{(N)}$ *and* $M''_{(N)}$*. If* $M'_{(N)}$ *and* $M''_{(N)}$ *are positive semidefinite, then so is* $M_{(N)}$.

Let us now prove the quantum Bochner theorem:

Theorem 165 (Quantum Bochner). *Let* $\widehat{A} = \operatorname{Op}_W(a)$ *be a self-adjoint trace-class operator on* $L^2(\mathbb{R}^n)$ *with symbol* $a \in L^1(\mathbb{R}^{2n})$*. We have* $\widehat{A} \geq 0$ *if and only if* $a_\diamond$ *is of ℏ-positive type.*

Proof. Notice that the self-adjointness of $\widehat{A}$ implies that the symbol a is real. If $\widehat{A} \geq 0$, then we have (formula (12.18))

$$a = (2\pi\hbar)^n \sum_j \alpha_j W\psi_j \tag{13.9}$$

for an orthonormal basis $(\psi_j)_j$ in $L^2(\mathbb{R}^n)$, the coefficients α_j being ≥ 0 and $(\alpha_j)_j \in \ell^1(\mathbb{N})$. It is thus sufficient to show that the Wigner transform $W\psi$ of an arbitrary $\psi \in L^2(\mathbb{R}^n)$ is of ℏ-positive type. This amounts to showing that, for all $(z_1, \dots, z_N) \in (\mathbb{R}^{2n})^N$ and all $(\zeta_1, \dots, \zeta_N) \in \mathbb{C}^N$, we have

$$I_N(\psi) = \sum_{1\leq j,k\leq N} \zeta_j \overline{\zeta_k} e^{-\frac{i}{2\hbar}\sigma(z_j,z_k)} F_\sigma W\psi(z_j - z_k) \geq 0 \tag{13.10}$$

for every complex vector $(\zeta_1, \dots, \zeta_N) \in \mathbb{C}^N$ and every sequence $(z_1, \dots, z_N) \in (\mathbb{R}^{2n})^N$. Since the Wigner distribution $W\psi$ and the ambiguity function $A\psi$ are obtained from each other by the symplectic Fourier transform F_σ, we have

$$I_N(\psi) = \sum_{1\leq j,k\leq N} \zeta_j \overline{\zeta_k} e^{-\frac{i}{2\hbar}\sigma(z_j,z_k)} A\psi(z_j - z_k).$$

Let us prove that

$$I_N(\psi) = \left(\frac{1}{2\pi\hbar}\right)^n \left\| \sum_{1\leq j\leq N} \zeta_j \widehat{D}(-z_j)\psi \right\|_{L^2}^2; \tag{13.11}$$

the inequality (13.10) will follow. Taking into account the fact that $\widehat{D}(-z_k)^* = \widehat{D}(z_k)$ and using the relationship (2.6), we have

$$\widehat{D}(z_k - z_j) = e^{-\frac{i}{2\hbar}\sigma(z_j,z_k)}\widehat{D}(z_k)\widehat{D}(-z_j), \tag{13.12}$$

hence, expanding the square in the right-hand side of (13.11),

$$\begin{aligned}
\left\| \sum_{1\leq j\leq N} \zeta_j \widehat{D}(-z_j)\psi \right\|_{L^2}^2 &= \sum_{1\leq j,k\leq N} \zeta_j\overline{\zeta_k}(\widehat{D}(-z_j)\psi|\widehat{D}(-z_k)\psi)_{L^2} \\
&= \sum_{1\leq j,k\leq N} \zeta_j\overline{\zeta_k}(\widehat{D}(z_k)\widehat{D}(-z_j)\psi|\psi)_{L^2} \\
&= \sum_{1\leq j,k\leq N} \zeta_j\overline{\zeta_k}e^{\frac{i}{2\hbar}\sigma(z_j,z_k)}(\widehat{D}(z_k - z_j)\psi|\psi)_{L^2} \\
&= (2\pi\hbar)^n \sum_{1\leq j,k\leq N} \zeta_j\overline{\zeta_k}e^{\frac{i}{2\hbar}\sigma(z_j,z_k)}A\psi(z_j - z_k),
\end{aligned}$$

and this proves the equality (13.11). Assuming that, conversely, $a_\diamond$ is of *ħ-positive type*, let us show that $(\widehat{A}\psi|\psi)_{L^2} \geq 0$ for all $\psi \in L^2(\mathbb{R}^n)$. Since the operator $\widehat{A}$ is bounded on $L^2(\mathbb{R}^n)$, it is sufficient to prove this for $\psi \in \mathcal{S}(\mathbb{R}^n)$, or equivalently

$$\int_{\mathbb{R}^{2n}} a(z)W\psi(z)dz \geq 0 \tag{13.13}$$

for $\psi \in \mathcal{S}(\mathbb{R}^n)$. Let us now write

$$M_{jk} = A\psi(z_k - z_j)a_\sigma(z_j - z_k);$$

we claim that the matrix $M_{(N)} = (M_{jk})_{1\leq j,k\leq N}$ is positive semidefinite. In fact, M is the Hadamard product of the positive semidefinite matrices $M'_{(N)} = (M'_{jk})_{1\leq j,k\leq N}$ and $M''_{(N)} = (M''_{jk})_{1\leq j,k\leq N}$ where

$$\begin{aligned}
M'_{jk} &= e^{\frac{i}{2\hbar}\sigma(z_j,z_k)}A\psi(-(z_j - z_k)) \\
M''_{jk} &= e^{-\frac{i}{2}\hbar\sigma(z_j,z_k)}a_\sigma(z_j - z_k)
\end{aligned}$$

and Lemma 164 implies that $M_{(N)}$ is also positive semidefinite. Hence, for the function b defined by

$$b_\sigma(z) = A\psi(-z)a_\sigma(z),$$

we have that $b_\diamond$ is of positive type. In particular, $|b_\sigma(z)| \leq b_\sigma(0)$, which implies $b_\sigma(0) > 0$ or $b_\sigma(z) = 0$ for every z. Moreover, $a_\sigma \in FL^1(\mathbb{R}^{2n})$ and $A\psi \in \mathcal{S}(\mathbb{R}^{2n})$, which implies that $b_\sigma \in L^1(\mathbb{R}^{2n}) \cap FL^1(\mathbb{R}^{2n})$, and similarly for b itself. If $b_\sigma(z)(0) > 0$, it follows from Bochner's theorem that b is, up to a positive constant, a probability density, hence $b(z) \geq 0$ for almost every $z \in \mathbb{R}^{2n}$, and therefore for every $z \in \mathbb{R}^{2n}$, because b is continuous. The same holds of course if $b_\sigma(z) = 0$ for every z. In either case, integrating the equality, we get, using the fact that a is real (because $\widehat{A}$ is self-adjoint) and the Plancherel formula for the symplectic Fourier transform,

$$\begin{aligned}(2\pi\hbar)^n b(0) &= \int_{\mathbb{R}^{2n}} A\psi(-z)a_\sigma(z)dz \\ &= \int_{\mathbb{R}^{2n}} A\psi(z)a_\sigma(-z)dz \\ &= \int_{\mathbb{R}^{2n}} A\psi(z)\overline{a_\sigma(z)}dz \\ &= \int_{\mathbb{R}^{2n}} W\psi(z)a(z)dz,\end{aligned}$$

hence the inequality (13.13) since $b(0) \geq 0$, proving the equality (13.11). □

13.4 The case of Gaussian distributions

Let Σ be a positive symmetric (real) $2n \times 2n$ matrix and consider the Gaussian

$$\rho(z) = \left(\frac{1}{2\pi}\right)^n \frac{1}{\sqrt{\det \Sigma}} e^{-\frac{1}{2}\Sigma^{-1}z^2}. \tag{13.14}$$

Let us find for which matrices Σ the function ρ is of $\hbar$-positive type.

13.4.1 Two lemmas

We begin by stating the multidimensional Hardy's uncertainty principle:

Lemma 166. *Let A and B be two real positive definite matrices and $\psi \in L^2(\mathbb{R}^n)$, $\psi \neq 0$. Assume that*

$$|\psi(x)| \leq Ce^{-\frac{1}{2}Ax^2} \quad \text{and} \quad |F\psi(p)| \leq Ce^{-\frac{1}{2}Bp^2} \tag{13.15}$$

for a constant $C > 0$. Then: (i) *The eigenvalues λ_j, $j = 1, \dots, n$, of the matrix AB are all $\leq 1/\hbar^2$; and* (ii) *if $\lambda_j = 1/\hbar^2$ for all j, then $\psi(x) = Ce^{-\frac{1}{2}Ax^2}$ for some constant C.*

We will also need the two following lemmas:

Lemma 167. *If R is a symmetric positive semidefinite $2n \times 2n$ matrix, then*

$$P_{(N)} = (Rz_j \cdot z_k)_{1 \le j,k \le N} \tag{13.16}$$

is a symmetric positive semidefinite $N \times N$ matrix for all $z_1, \dots, z_N \in \mathbb{R}^{2n}$.

Proof. There exists a matrix L such that $R = L^* L$ (Cholesky decomposition). Denoting by $\langle z|z' \rangle = z \cdot \overline{z'}$ the inner product on $\mathbb{C}^{2n}$, we have, since the z_j are real vectors,

$$L^* z_j \cdot z_k = \langle L^* z_j | z_k \rangle = \langle z_j | L z_k \rangle = z_j \cdot \overline{Lz_k}$$

hence $Rz_j \cdot z_k = Lz_j \cdot \overline{Lz_k}$. It follows that for all complex ζ_j we have

$$\sum_{1 \le j,k \le N} \zeta_j \overline{\zeta_k} R z_j \cdot z_k = \sum_{1 \le j \le N} \zeta_j L z_j \overline{\left(\sum_{1 \le j \le N} \zeta_j L z_j \right)} \ge 0$$

hence our claim. □

13.4.2 The condition on Σ

We now have the tools needed to give a complete characterization of the $\hbar$-positivity of Gaussians (13.14). Recall that the symplectic eigenvalues of a positive definite symmetric matrix Σ are the numbers $\lambda_j > 0$ such that $\pm i\lambda_j$ is an eigenvalue of $J\Sigma$. One has Williamson's symplectic diagonalization result $M = S^T D S$ where D is a diagonal matrix of the type

$$D = \begin{pmatrix} \Lambda & 0 \\ 0 & \Lambda \end{pmatrix},$$

the diagonal entries of Λ being the positive numbers λ_j (see Chapter 1).

Proposition 168. *The Gaussian function* (13.14) *is $\hbar$-positive if and only if*

$$\hbar \le 2\lambda_{\min} \tag{13.17}$$

where $\lambda_{\min}$ is the smallest symplectic eigenvalue of Σ; equivalently

$$\Sigma + \frac{i\hbar}{2} J \ge 0. \tag{13.18}$$

Proof. Let us first show that the conditions (13.17) and (13.18) are equivalent. Let $\Sigma = S^T D S$ be a symplectic diagonalization of Σ. Since $S^T J S = J$ condition (13.18) is equivalent to $D + \frac{i\hbar}{2} J \ge 0$, let $z = (x, p)$ be an eigenvector of $D + \frac{i\hbar}{2} J$; the corresponding eigenvalue λ is real and ≥ 0. The equation $(D + \frac{i\hbar}{2}J)z = \lambda z$ is easily seen to be equivalent to the equations $(\Lambda - \lambda)^2 x = \frac{1}{4}\hbar^2 x$ and $(\Lambda - \lambda)^2 p = \frac{1}{4}\hbar^2 p$. Since $z \neq 0$, we must have $(\lambda_j - \lambda)^2 = \frac{1}{4}\hbar^2$, that is $\lambda_j - \lambda = \pm\frac{1}{2}\hbar$ for all $j = 1, 2, \ldots, n$; since $\lambda \geq 0$, this is equivalent to $\lambda_j \geq \frac{1}{2}\hbar$ for all j, that is to (13.17). Let us now show that the condition (13.17) is necessary for the function

$$a(z) = (2\pi)^{-n}\sqrt{\det \Sigma^{-1}}e^{-\frac{1}{2}\Sigma^{-1}z^2} \tag{13.19}$$

to be a $\hbar$-Wigner transform of a positive trace class operator. Let $\widehat{A} = \mathrm{Op_W}(a)$ and $\widehat{S} \in \mathrm{Mp}(n)$; the operator $\widehat{A}$ is of trace class if and only if $\widehat{S}\widehat{A}\widehat{S}^{-1}$ is, in which case $\mathrm{Tr}(\widehat{A}) = \mathrm{Tr}(\widehat{S}\widehat{A}\widehat{S}^{-1})$. Choose $\widehat{S}$ with projection $S \in \mathrm{Sp}(n)$ such that $\Sigma = S^TDS$ is a symplectic diagonalization of Σ. This choice reduces the proof to the case $\Sigma = D$, that is, to

$$a(z) = (2\pi)^{-n}(\det \Lambda^{-1})e^{-\frac{1}{2}(\Lambda^{-1}x^2+\Lambda^{-1}p^2)}. \tag{13.20}$$

Suppose now that $\widehat{A}$ is of trace class; then by the spectral theorem (Proposition 145), there exists an orthonormal basis of functions $(\psi_j)_j$ of $L^2(\mathbb{R}^n)$ such that

$$\rho(z) = \sum_j \alpha_j W\psi_j(z)$$

where the α_j are real numbers summing up to one. Integrating with respect to the p and x variables, respectively, the marginal conditions satisfied by the Wigner transform imply that we have

$$\sum_j \alpha_j |\psi_j(x)|^2 = (2\pi)^{-n/2}(\det \Lambda)^{1/2}e^{-\frac{1}{2}\Lambda^{-1}x^2}$$

$$\sum_j \alpha_j |F\psi_j(p)|^2 = (2\pi)^{-n/2}(\det \Lambda)^{1/2}e^{-\frac{1}{2}\Lambda^{-1}p^2}.$$

In particular, since $\alpha_j \geq 0$ for every $j = 1, 2, \ldots, n$,

$$|\psi_j(x)| \leq C_j e^{-\frac{1}{4}\Lambda^{-1}x^2}, \quad |F\psi_j(p)| \leq C_j e^{-\frac{1}{4}\Lambda^{-1}p^2}$$

with $C_j = (2\pi)^{-n/4}(\det \Lambda)^{1/4}/\alpha_j^{1/2}$. Applying Lemma 166 with $A = B = \frac{\hbar}{2}\Lambda^{-1}$, we must have $\hbar \leq 2\lambda_j$ for all $j = 1, \ldots, n$, which is condition (13.17); this establishes the sufficiency statement. (iii) Let us finally show that, conversely, the condition (13.18) is sufficient. It is again no restriction to assume that Σ is the diagonal matrix $D = \begin{pmatrix} \Lambda & 0 \\ 0 & \Lambda \end{pmatrix}$; the symplectic Fourier transform of ρ is easily calculated, and one finds that $\rho_\diamond(z) = e^{-\frac{1}{4}Dz^2}$. Let $\Lambda_{(N)} = (\Lambda_{jk})_{1\leq j,k\leq N}$ with

$$\Lambda_{jk} = e^{-\frac{i\hbar}{2}\sigma(z_j,z_k)}\rho_\diamond(z_j - z_k);$$

a simple algebraic calculation shows that we have

$$\Lambda_{jk} = e^{-\frac{1}{4}Dz_j^2}e^{\frac{1}{2}(D+i\hbar J)z_j\cdot z_k}e^{-\frac{1}{4}Dz_k^2},$$

and hence

$$\Lambda_{(N)} = \Delta_{(N)}\Gamma_{(N)}\Delta^*_{(N)}$$

where $\Delta_{(N)} = \operatorname{diag}(e^{-\frac{1}{4}Dz_1^2}, \ldots, e^{-\frac{1}{4}Dz_N^2})$ and $\Gamma_{(N)} = (\Gamma_{jk})_{1\leq j,k\leq N}$ with $\Gamma_{jk} = e^{\frac{1}{2}(D+i\hbar J)z_j\cdot z_k}$. The matrix $\Lambda_{(N)}$ is thus positive semidefinite if and only if $\Gamma_{(N)}$ is, but this is the case in view of Lemma 167. □

13.5 Positive trace-class operators

We are now going to give an integral description of the positivity of a trace-class operator on $L^2(\mathbb{R}^n)$. We begin with a general result:

Proposition 169. *Let* $\widehat{A} = \operatorname{Op_W}(a)$ *be a trace-class operator on* $L^2(\mathbb{R}^n)$*. We have* $\widehat{A} \geq 0$ *if and only if* $\operatorname{Tr}(\widehat{A}\widehat{B}) \geq 0$ *for every positive operator* $\widehat{B} \in \mathcal{L}_1(L^2(\mathbb{R}^n))$.

Proof. Since $\mathcal{L}_1(L^2(\mathbb{R}^n))$ is an algebra the product, $\widehat{A}\widehat{B}$ is indeed of trace class so the condition $\operatorname{Tr}(\widehat{A}\widehat{B}) \geq 0$ makes sense; setting $\widehat{B} = \operatorname{Op_W}(b)$, we have

$$b(z) = (2\pi\hbar)^n \sum_j \beta_j W\psi_j$$

where $(\beta_j)_j \in \ell^1(\mathbb{N})$ with $\beta_j \geq 0$ and ψ_j an orthonormal basis for $L^2(\mathbb{R}^n)$. Observing that trace-class operators are also Hilbert–Schmidt ones, we also have, since $a, b \in L^2(\mathbb{R}^{2n})$,

$$\operatorname{Tr}(\widehat{A}\widehat{B}) = \int_{\mathbb{R}^{2n}} a(z)b(z)dz, \tag{13.21}$$

and hence

$$\operatorname{Tr}(\widehat{A}\widehat{B}) = (2\pi\hbar)^n \sum_j \beta_j \int_{\mathbb{R}^{2n}} a(z)W\psi_j(z)dz$$

(the interchange of integral and series is justified by Fubini's Theorem). Assume that $\operatorname{Tr}(\widehat{A}\widehat{B}) \geq 0$. It is enough to check the positivity of $\widehat{A}$ on unit vectors ψ in $L^2(\mathbb{R}^n)$. Choosing all the $\beta_j = 0$ except β_1 and setting $\psi_1 = \psi$, we have

$$\int_{\mathbb{R}^{2n}} a(z)W\psi(z)dz \geq 0; \tag{13.22}$$

since we can choose $\psi \in L^2(\mathbb{R}^n)$ arbitrarily, this means that we have $\widehat{A} \geq 0$. If, conversely, we have $\widehat{A} \geq 0$, then (13.22) holds for all ψ_j, hence $\operatorname{Tr}(\widehat{A}\widehat{B}) \geq 0$. □

Let us next prove:

Proposition 170. *Let $\widehat{A} = \mathrm{Op_W}(a)$ be a trace-class operator. We have $\widehat{A} \geq 0$ if and only if*

$$\int_{\mathbb{R}^{2n}} F_\sigma a(z)\Big(\int_{\mathbb{R}^{2n}} e^{-\frac{i}{2\hbar}\sigma(z,z')}c(z'-z)\overline{c(z')}dz'\Big)dz \geq 0 \tag{13.23}$$

for all $c \in L^2(\mathbb{R}^{2n})$.

Proof. In view of Lemma 169, we have $\widehat{A} \geq 0$ if and only if $\mathrm{Tr}(\widehat{A}\widehat{B}) \geq 0$ for all positive $\widehat{B} = \mathrm{Op_W}(b) \in \mathcal{L}_1(L^2(\mathbb{R}^n))$, that is, (formula (13.21))

$$\int_{\mathbb{R}^{2n}} a(z)b(z)dz \geq 0 = (a|b)_{L^2(\mathbb{R}^{2n})}. \tag{13.24}$$

(Recall that Weyl symbols of self-adjoint operators are real). Using Plancherel's Theorem,

$$(a|b)_{L^2(\mathbb{R}^{2n})} = \left(\frac{1}{2\pi\hbar}\right)^n (F_\sigma a|F_\sigma b)_{L^2(\mathbb{R}^{2n})}. \tag{13.25}$$

Since $\widehat{B} \geq 0$, there exists $\widehat{C} \in \mathcal{L}_2(L^2(\mathbb{R}^n))$ such that $\widehat{B} = \widehat{C}^*\widehat{C}$, and hence, setting $\widehat{C} = \mathrm{Op_W}(c)$ (recall that $\widehat{C}^* = \mathrm{Op_W}(\bar{c})$), by the composition formula for Weyl operators

$$\begin{aligned} F_{\sigma,\hbar}b(z) &= \left(\frac{1}{2\pi\hbar}\right)^n \int_{\mathbb{R}^{2n}} e^{\frac{i}{2\hbar}\sigma(z,z')}F_\sigma\bar{c}(z-z')F_\sigma c(z')dz' \\ &= \left(\frac{1}{2\pi\hbar}\right)^n \int_{\mathbb{R}^{2n}} e^{\frac{i}{2\hbar}\sigma(z,z')}\overline{F_\sigma c}(z'-z)F_{\sigma,\hbar}c(z')dz'. \end{aligned}$$

Vice-versa, taking any function $c \in L^2(\mathbb{R}^{2n})$, then $\widehat{C} = \mathrm{Op_W}(c)$ is a Hilbert–Schmidt operator, and $\widehat{B} = \widehat{C}^*\widehat{C}$ is a positive operator. Hence, using that the operator $F_{\sigma,\hbar}$ is a topological isomorphism on $L^2(\mathbb{R}^{2n})$, condition (13.24) and (13.25) are equivalent to

$$\int_{\mathbb{R}^{2n}} F_\sigma a(z)\Big(\int_{\mathbb{R}^{2n}} e^{-\frac{i}{2\hbar}\sigma(z,z')}c(z'-z)\overline{c(z')}dz'\Big)dz \geq 0, \tag{13.26}$$

for every $c \in L^2(\mathbb{R}^{2n})$, as we set out to prove. □

13.6 Comments and references

The topic of positivity, which was quite active in the late 1980's, hasn't much evolved since; many open questions remain open. It is a notoriously difficult part of functional analysis that has been tackled by many authors, but there have been few decisive advances since the pioneering work of Kastler [50] and Loupias and Miracle-Sole [55, 56]; see however Dias and Prata [18] and our joint paper [17] with Cordero and Nicola.

For a proof of Bochner's theorem, see Katznelson [51]. The proof of the quantum Bochner theorem goes back to the seminal work of Kastler [50] and Loupias and Miracle-Sole [55, 56]. While these authors use the theory of C^*-algebras and hard functional analysis, the conceptually simpler proof we have given here was published in our joint work with Elena Cordero and Fabio Nicola [17] already mentioned. In the literature, the conditions (13.8) in Definition 161 are often called the "KLM conditions" as a tribute to Kastler, Loupias, and Miracle-Sole. (Theorem 4). Narcowich [58] was the first to study the positivity of Gaussians using the approach in Kastler's paper [50]; the proof we have given in Proposition 168 is simpler. For a proof of Proposition 166, see de Gosson and Luef [37].

14 The density operator

Quantum mechanics in its modern form is the study of density operators and of their properties, both geometric and analytical. The density operator (or density matrix as it is called in physics) describes a statistical ensemble of pure quantum states. Its theory is far from being complete, for instance, the study of entangled mixed states (which we study later on in this book) contains many open problems.

14.1 Mixed quantum states: heuristics

A quantum state is, mathematically speaking, an operator that provides a probability distribution for the outcomes of each possible measurement on a system. A quantum system is said to be in a *pure* state if we have complete knowledge about that system. Physicists usually denote a pure state by $|\psi\rangle$ (Dirac's ket notation). Mathematically, the state is identified with the linear span $\mathbb{C}\psi = \{\lambda\psi : \lambda \in \mathbb{C}\}$ of a non-zero function ψ belonging to some Hilbert space H that we will usually identify with $L^2(\mathbb{R}^n)$. On a slightly more abstract level, this pure state can be identified with the orthogonal projection $\widehat{\rho}_\psi$ on the ray $\mathbb{C}\psi$. This projection, which is of rank one, is denoted by $|\psi\rangle\langle\psi|$ in quantum mechanics. However, in practice we have only partial knowledge of the quantum system under consideration. When this is the case, we say that the system is in a *mixed state*. Mixed states are classical probabilistic mixtures of pure states; however, different distributions of pure states can generate physically indistinguishable mixed states. Technically speaking, a quantum mixed state can be viewed as the datum of a set of pairs $\{(\psi_j, \lambda_j)\}$ where ψ_j is a (normalized square integrable) pure state and λ_j $(0 \leq \lambda_j \leq 1)$ a classical probability, these probabilities summing up to one: $\sum_j \lambda_j = 1$. One identifies the state $\{(\psi_j, \lambda_j)\}$ with the convex sum of projectors

$$\widehat{\rho} = \sum_j \lambda_j \widehat{\rho}_j$$

where $\widehat{\rho}_j = \widehat{\rho}_{\psi_j}$ is the orthogonal projection on the ray $\mathbb{C}\psi_j$. The operator $\widehat{\rho}$ is called a *density operator* (or density matrix in physics). It is a trace class operator with trace $\operatorname{Tr}(\widehat{\rho}) = 1$. The *Wigner distribution* of the density operator $\widehat{\rho}$ (or of the state $\{(\psi_j, \lambda_j)\}$) is the function

$$\rho = \sum_j \lambda_j W\psi_j,$$

and this function plays the role of a probability distribution on $\mathbb{R}^n$ (although it can take negative values): The density operator describes a preparation procedure for an ensemble of quantum systems whose statistical properties correspond to the given preparation procedure

https://doi.org/10.1515/9783110722772-014

14.2 Density operator: definition and first properties

14.2.1 Definition of a density operator

Recapitulating from our informal discussion:

Definition 171. A quantum state is a set of pairs $(\psi_j, \lambda_j) \in L^2 \times \mathbb{R}_+$ indexed by a discrete set F and where $\|\psi_j\|_{L^2} = 1$ for all $j \in F$ and $\sum_{j\in F} \lambda_j = 1$. The linear operator $\widehat{\rho}$ on $L^2(\mathbb{R}^n)$ defined by

$$\widehat{\rho}\psi = \sum_{j\in F} \lambda_j (\psi|\psi_j)_{L^2} \psi_j \tag{14.1}$$

is the density operator associated with that state; the Wigner distribution of that state is

$$\rho = \sum_{j\in F} \lambda_j W\psi_j. \tag{14.2}$$

It is customary in quantum mechanics to identify a quantum state with the density operator it determines.

Definition (14.1) can be rewritten

$$\widehat{\rho} = \sum_{j\in F} \lambda_j \widehat{\rho}_j$$

where $\widehat{\rho}_j$ is the orthogonal projector on the ray $\mathbb{C}\psi_j$. Notice that this is absolutely convergent in $\mathcal{B}(L^2(\mathbb{R}^n))$ since $\|\widehat{\rho}_j\|_{\mathcal{B}(L^2(\mathbb{R}^n))} = 1$ for every j and

$$\sum_j \|\lambda_j \widehat{\rho}_j\|_{\mathcal{B}(L^2(\mathbb{R}^n))} \leq \sum_j |\lambda_j| < \infty.$$

We will usually skip the reference to the index set F and write simply

$$\widehat{\rho} = \sum_j \lambda_j \widehat{\rho}_j.$$

14.2.2 First properties

The first (hardly surprising) result is:

Proposition 172. *The operator $\widehat{\rho} = \sum_j \lambda_j \widehat{\rho}_j$ is a self-adjoint semidefinite positive operator on $L^2(\mathbb{R}^n)$, and we have $\widehat{\rho} \in \mathcal{L}_1(L^2(\mathbb{R}^n))$ and $\operatorname{Tr}(\widehat{\rho}) = 1$.*

Proof. That $\widehat{\rho} \geq 0$ follows from the fact that (formula (14.1))

$$(\widehat{\rho}\psi|\psi)_{L^2} = \sum_j \lambda_j (\psi|\psi_j)_{L^2} (\psi_j|\psi)_{L^2}$$

$$= \sum_j \lambda_j |(\psi|\psi_j)_{L^2}|^2 \geq 0.$$

Notice that it follows from the Cauchy–Schwarz inequality that the series is absolutely convergent (it is also easy to verify directly) and that

$$(\widehat{\rho}\psi|\psi)_{L^2} \leq \|\psi\|^2_{L^2}.$$

The self-adjointness is clear since we are dealing with a complex Hilbert space, so it remains to be shown that $\widehat{\rho}$ is of trace class and $\mathrm{Tr}(\widehat{\rho}) = 1$. Let $(\phi_k)_k$ be an orthonormal basis of $L^2(\mathbb{R}^n)$;, we have $\widehat{\rho}_j\phi_k = (\phi_k|\psi_j)_{L^2}\psi_j$, hence $(\widehat{\rho}_j\phi_k|\phi_k)_{L^2} = |(\phi_k|\psi_j)_{L^2}|^2$, and hence, using Parseval's identity,

$$\begin{aligned}\sum_k (\widehat{\rho}\phi_k|\phi_k)_{L^2} &= \sum_{j,k} \lambda_j |(\phi_k|\psi_j)_{L^2}|^2 \\ &= \sum_j \lambda_j \Big(\sum_k |(\phi_k|\psi_j)_{L^2}|^2 \Big) \\ &= \sum_j \lambda_j \|\psi_j\|^2_{L^2} < \infty,\end{aligned}$$

hence $\widehat{\rho}$ is of trace class, and $\mathrm{Tr}(\widehat{\rho}) = \sum_j \lambda_j = 1$ since $\|\psi_j\|_{L^2} = 1$ for all j. □

For the Wigner distribution, we have:

Proposition 173. *The series $\rho = \sum_j \lambda_j W\psi_j$, $\lambda_j \geq 0$, $\sum_j \lambda_j = 1$ converges in $L^2(\mathbb{R}^{2n})$, and we have*

$$\rho \in L^2(\mathbb{R}^{2n}) \cap L^\infty(\mathbb{R}^{2n}).$$

In view of Moyal's identity (3.12), we have

$$\|\lambda_j W\psi_j\|_{L^2} = \lambda_j \left(\frac{1}{2\pi\hbar}\right)^{\frac{n}{2}} \|\psi_j\|^2_{L^2} = \left(\frac{1}{2\pi\hbar}\right)^{\frac{n}{2}} \lambda_j,$$

and hence $\|\rho\|_{L^2} \leq (2\pi\hbar)^{-n/2}$ since the λ_j are ≥ 0 and sum up to one. We therefore have absolute convergence in $L^2(\mathbb{R}^{2n})$ of the series ρ. On the other hand, by Cauchy–Schwarz's inequality (recall that $\|\psi_j\|^2_{L^2} = 1$),

$$|W\psi_j(z)| \leq \left(\frac{2}{\pi\hbar}\right)^n$$

for all $z \in \mathbb{R}^{2n}$, hence

$$\|\rho\|_\infty \leq \sum_j \lambda_j \|W\psi_j\|_\infty \leq \left(\frac{2}{\pi\hbar}\right)^n,$$

and the series ρ is absolutely convergent in $L^\infty(\mathbb{R}^{2n})$.

Remark 174. Notice that in general ρ is not in $L^1(\mathbb{R}^{2n})$. Let for instance $\rho = W\psi$; if $\rho \in L^1(\mathbb{R}^{2n})$, then ψ must be in the Feichtinger algebra $S_0(\mathbb{R}^{2n})$ (Proposition 117) which is smaller than $L^1(\mathbb{R}^n)$. We will come back to this in a moment.

14.2.3 The convexity property

Here is an interesting geometric result about the set of density operators.

Proposition 175. (i) *The density operators on a Hilbert space H form a convex subset* $\mathrm{Dens}(H)$ *of the space* $\mathcal{B}(H)$ *of bounded operators on H.* (ii) *Let $\widehat{\rho}_\psi$ be the orthogonal projection on the ray $\mathbb{C}\psi$ ($\|\psi\|_H = 1$). If there exist $\widehat{\rho}_1, \widehat{\rho}_2 \in \mathrm{Dens}(H)$ such that $\widehat{\rho}_\psi = \lambda\widehat{\rho}_1 + (1-\lambda)\widehat{\rho}_2$ with $\lambda \neq 0$ and $\lambda \neq 1$, then $\widehat{\rho}_\psi = \widehat{\rho}_1 = \widehat{\rho}_2$. That is, the extreme points of* $\mathrm{Dens}(H)$ *are the rank-one projections that correspond to the pure states on H.*

Proof. (i) To say that the set of density matrices is convex means that, if $\widehat{\rho}_1$ and $\widehat{\rho}_2$ are in $\mathrm{Dens}(H)$, then so is $\lambda\widehat{\rho}_1 + (1-\lambda)\widehat{\rho}_2$ for all real numbers λ such that $0 \leq \lambda \leq 1$. Let $\{(\psi_j, \alpha_j) : j \in F\}$ and $\{(\phi_j, \beta_j) : j \in G\}$ be two mixed states. Relabeling if necessary the indices, we may assume that the sets F and G are disjoint. The corresponding density matrices are

$$\widehat{\rho}_1 = \sum_{k\in F} \alpha_k \widehat{\rho}_{\psi_k}, \quad \widehat{\rho}_2 = \sum_{\ell\in G} \beta_\ell \widehat{\rho}_{\phi_\ell}$$

where $\widehat{\rho}_{\psi_k}$ and $\widehat{\rho}_{\phi_\ell}$ are the orthogonal projections on $\mathbb{C}\psi_k$ and $\mathbb{C}\phi_\ell$, respectively, and hence we have by linearity

$$\lambda\widehat{\rho}_1 + (1-\lambda)\widehat{\rho}_2 = \sum_{j\in F\cup G} \gamma_j \widehat{\rho}_{\chi_j}$$

where $\chi_j = \psi_j$, $\gamma_j = \lambda\alpha_j$ if $j \in F$ and $\chi_j = \phi_j$, $\gamma_j = (1-\lambda)\beta_j$ if $j \in G$. That $\widehat{\rho} = \lambda\widehat{\rho}_1 + (1-\lambda)\widehat{\rho}_2$ is a mixed state now follows from the obvious equality

$$\sum_{j\in F\cup G} \gamma_j = \lambda \sum_{j\in F} \alpha_j + (1-\lambda) \sum_{j\in G} \beta_j = 1.$$

(ii) That a pure state never can be represented as a mixed state is easily seen using the following algebraic argument: Assume that $\widehat{\rho}_\psi$ can be rewritten as a sum $\sum_j \alpha_j \widehat{\rho}_{\psi_j}$ with $\alpha_j \geq 0$ and $\sum_j \alpha_j = 1$. Discarding the terms with $\alpha_j = 0$, we may assume that $\alpha_j > 0$ for all indices j. Let now $(\mathbb{C}\psi)^\perp$ be the subspace of H orthogonal to the ray $\mathbb{C}\psi$: It consists of all vectors ϕ in H such that $(\psi|\phi)_{L^2} = 0$. For every $\phi \in (\mathbb{C}\psi)^\perp$, we have $(\widehat{\rho}_\psi\phi|\phi)_H = |(\phi|\psi)_H|^2 = 0$, and hence also, if we assume that $\widehat{\rho}_\psi = \sum_j \alpha_j \widehat{\rho}_{\psi_j}$,

$$(\widehat{\rho}_\psi\phi|\phi)_H = \sum_j \alpha_j |(\phi|\psi_j)_H|^2 = 0.$$

This equality implies that we must have $(\phi|\psi_j)_{L^2} = 0$ for every $\phi \in (\mathbb{C}\psi)^\perp$ and every ψ_j, hence $\psi_j \in ((\mathbb{C}\psi)^\perp)^\perp = \mathbb{C}\psi$. Since ψ and ψ_j have norms equal to one, we must have $\psi_j = \lambda_j\psi$ for some complex number λ_j with $|\lambda_j| = 1$; the vectors ψ_j thus define the state ψ. This means that, if $\widehat{\rho}_\psi$ is a pure state density matrix, then the relationship $\widehat{\rho}_\psi = \lambda\widehat{\rho}_1 + (1-\lambda)\widehat{\rho}_2$ with $\lambda \neq 0$ and $\lambda \neq 1$ implies $\widehat{\rho}_\psi = \widehat{\rho}_1 = \widehat{\rho}_2$. That this, is the case immediately follows from the previous argument. □

14.2.4 The spectral theorem for density operators

We now have most of the tools needed for a functional study of the density operator

$$\widehat{\rho}\psi = \sum_{j\in F} \lambda_j(\psi|\psi_j)_{L^2}\psi_j. \tag{14.3}$$

Let us now state the spectral theorem for density operators. It is just a slight reformulation of the result for arbitrary trace class operators.

Theorem 176 (Spectral Theorem). *Let $\widehat{\rho}$ be a density operator on $L^2(\mathbb{R}^n)$. Then:* (i) *The set (λ_j) of eigenvalues λ_j of $\widehat{\rho}$ is at most countable and consists of non-negative numbers $\lambda_1 \geq \lambda_2 \geq \cdots \geq \lambda_j \geq \cdots \geq 0$ such that $\sum_j \lambda_j = 1$, and we have $\lim_{j\to\infty} \lambda_j = 0$ if (λ_j) is infinite;* (ii) *the multiplicity of every eigenvalue λ_j is finite;* (iii) *there exists an at-most countable set (ψ_j) of orthonormal eigenvectors of $\widehat{A}$ with eigenvalues λ_j such that, if ψ_j is an eigenvector for λ_j, then*

$$\widehat{\rho}\psi = \sum_{j} \lambda_j(\psi|\psi_j)_{L^2}\psi_j. \tag{14.4}$$

In particular, the Wigner distribution of $\widehat{\rho}$ is the sum

$$\rho = \sum_j \lambda_j W\psi_j \tag{14.5}$$

of pairwise orthogonal Wigner transforms $W\psi_j$,

Proof. It is an immediate consequence of the spectral theorem for compact operators (Theorem 7). That the eigenvalues are ≥ 0 follows from the positive semi-definiteness of $\widehat{\rho}$, and the relationship $\sum_j \lambda_j = 1$ follows from $\mathrm{Tr}(\widehat{\rho}) = 1$. That the $W\psi_j$ are pairwise orthogonal follows from Moyal's identity. □

This result makes clear that different mixed states can give rise to the same operator: In our original definition (Definition 171) of the density matrix, there was no orthogonality requirement for the pure state composing the mixed state; the theorem shows that one can always find a system of orthonormal states defining that operator. The physical significance of this ambiguity is not well-understood and deserves further study. Let us just say that the statistical information we have about a mixed state is encoded in the density operator, so that the predictions we can make about the state are not dependent on the way it is decomposed.

The spectral theorem allows us to study the purity of a quantum state that is defined as follows:

Definition 177. Let $\widehat{\rho}$ be a density operator on $L^2(\mathbb{R}^n)$. The purity of $\widehat{\rho}$ is the positive real number

$$\mu(\widehat{\rho}) = \mathrm{Tr}(\widehat{\rho}^2). \tag{14.6}$$

The definition makes sense because $\widehat{\rho}^2 \in \mathcal{L}_1 L(^2(\mathbb{R}^n))$ in view of the algebra property of the trace class. Purity is a measure of the "mixedness" of a quantum state:

Proposition 178. (i) *We have*

$$\mu(\widehat{\rho}) = \sum_j \lambda_j^2 \tag{14.7}$$

where the λ_j are the eigenvalues of $\widehat{\rho}$. (ii) *We have $0 \leq \mu(\widehat{\rho}) \leq 1$ and $\mu(\widehat{\rho}) = 1$ if and only if $\widehat{\rho}$ is a pure state, that is, if there exists $\psi \in L^2(\mathbb{R}^n)$ such that $\widehat{\rho} = \widehat{\rho}_\psi$, the orthogonal projection on the ray $\mathbb{C}\psi$.*

Proof. (i) Since $\widehat{\rho} = (2\pi\hbar)^n \operatorname{Op_W}(\rho)$, the trace formula (12.16) yields

$$\operatorname{Tr}(\widehat{\rho}^2) = (2\pi\hbar)^n \int_{\mathbb{R}^{2n}} \rho(z)^2 dz. \tag{14.8}$$

Writing ρ as the sum (14.5) of mutually orthogonal Wigner transforms, we have

$$\rho^2 = \sum_j \lambda_j^2 (W\psi_j)^2,$$

hence

$$\operatorname{Tr}(\widehat{\rho}^2) = (2\pi\hbar)^n \sum_j \lambda_j^2 \int_{\mathbb{R}^{2n}} (W\psi_j(z))^2 dz.$$

In view of the Moyal identity (3.14) for the Wigner transform, we have

$$\int_{\mathbb{R}^{2n}} (W\psi_j(z))^2 dz = \left(\frac{1}{2\pi\hbar}\right)^n \|\psi_j\|_{L^2}^2 = \left(\frac{1}{2\pi\hbar}\right)^n,$$

hence formula (14.7). (ii) Since $\sum_j \lambda_j = 1$ and $\lambda_j \geq 0$, we have $\mu(\widehat{\rho}) = \sum_j \lambda_j^2 \leq 1$. The equality $\mu(\widehat{\rho}) = 1$ can occur only if all the λ_j except one are equal to zero. □

We will see later on that the purity of a Gaussian state with the Wigner distribution

$$\rho(z) = \left(\frac{1}{2\pi}\right)^n (\det \Sigma)^{-1/2} e^{-\frac{1}{2}\Sigma^{-1} z \cdot z}$$

is given by

$$\mu(\widehat{\rho}) = \left(\frac{\hbar}{2}\right)^n \det(\Sigma^{-1/2}).$$

14.3 The Wigner distribution as a Weyl symbol

14.3.1 The density operator as a Weyl operator

While the definition of the Wigner distribution of a density operator seems at first sight to be *ad hoc*, it is canonically related to $\widehat{\rho}$ by the Weyl transform:

Proposition 179. *The density operator $\widehat{\rho}$ can be expressed in terms of its Wigner distribution ρ by the formula*

$$\widehat{\rho} = (2\pi\hbar)^n \operatorname{Op_W}(\rho) . \tag{14.9}$$

Equivalently, using the displacement and reflection operators $\widehat{D}(z)$ and $\widehat{R}(z)$, we have the formulas

$$\widehat{\rho}\psi = 2^n \int_{\mathbb{R}^{2n}} \rho(z)\widehat{R}(z)\psi dz \tag{14.10}$$

and

$$\widehat{\rho}\psi = \int_{\mathbb{R}^{2n}} \rho_\sigma(z)\widehat{D}(z)\psi dz, \tag{14.11}$$

valid for $\psi \in L^2(\mathbb{R}^n)$.

Proof. Let $(\psi, \phi) \in \mathcal{S}(\mathbb{R}^n) \times \mathcal{S}(\mathbb{R}^n)$. We have

$$\begin{aligned}(\widehat{\rho}\psi|\phi)_{L^2} &= \sum_j \lambda_j(\psi|\psi_j)_{L^2}(\psi_j|\phi)_{L^2} \\ &= (2\pi\hbar)^n \sum_j \lambda_j W(\psi,\phi)W\psi_j\end{aligned}$$

where the second equality follows from Moyal's identity (3.12). In view of the definition (3.1) of $W(\psi, \phi)$, this can be rewritten

$$\begin{aligned}(\widehat{\rho}\psi|\phi)_{L^2} &= 2^n \int_{\mathbb{R}^{2n}} (\psi|\widehat{R}(z)\phi)_{L^2}\rho(z)dz \\ &= 2^n \int_{\mathbb{R}^{2n}} \rho(z)(\widehat{R}(z)\psi|\phi)_{L^2}dz\end{aligned}$$

where we have used the fact that the reflection operator. $\widehat{R}(z)$ is involutive. Formula (14.10) follows. To prove formula (14.11), we first note that (14.10) can be rewritten, using Plancherel's formula, as

$$\widehat{\rho}\psi = 2^n \int_{\mathbb{R}^{2n}} F_\sigma\rho(z)F_\sigma(\widehat{R}(\cdot)\psi)(-z)dz.$$

Recalling (formula (2.19) in Proposition 26) that

$$\widehat{R}(z)\psi = 2^{-n}F_\sigma[\widehat{D}(\cdot)\psi](-z),$$

we also have, since F_σ is an involution of $\mathcal{S}(\mathbb{R}^{2n})$,

$$F_\sigma(\widehat{R}(\cdot)\psi)(z) = 2^n\widehat{D}(-z)\psi$$

hence formula (14.11). □

14.3.2 Application to the statistical interpretation of $\widehat{\rho}$

We have seen in Section 3.5 that the Wigner transform of a normalized function $\psi \in L^2(\mathbb{R}^n)$ can be viewed as a quasi-probability function for pure states. The generalization to the density operator is rather straightforward.

Let $\widehat{A}$ be a self-adjoint operator on $L^2(\mathbb{R}^n)$; we assume for convenience that $\widehat{A}$ is bounded: $\widehat{A} \in \mathcal{B}(L^2(\mathbb{R}^n))$. We will refer to $\widehat{A}$ as an “observable”.

Definition 180. Let $\widehat{\rho}$ be a density operator on $L^2(\mathbb{R}^n)$. The average (or mean) value of the observable $\widehat{A}$ in the state $\widehat{\rho}$ is the real number

$$\langle\widehat{A}\rangle_{\widehat{\rho}} = \mathrm{Tr}(\widehat{A}\widehat{\rho}). \tag{14.12}$$

The number $\langle\widehat{A}\rangle_{\widehat{\rho}}$ is well-defined: Recall that the trace class operators form an ideal $\mathcal{L}_1(L^2(\mathbb{R}^n))$ in $\mathcal{B}(L^2(\mathbb{R}^n))$, hence $\widehat{A}\widehat{\rho} \in \mathcal{L}_1(L^2(\mathbb{R}^n))$. That $\mathrm{Tr}(\widehat{A}\widehat{\rho})$ is real follows from the observation that the trace of a self-adjoint operator is real and that

$$\overline{\mathrm{Tr}(\widehat{A}\widehat{\rho})} = \mathrm{Tr}((\widehat{A}\widehat{\rho})^*) = \mathrm{Tr}(\widehat{\rho}\widehat{A}) = \mathrm{Tr}(\widehat{A}\widehat{\rho}).$$

The following result connects the definition above to the Wigner formalism:

Proposition 181. *Assume that* $\widehat{A} = \mathrm{Op_W}(a)$. *We have*

$$\langle\widehat{A}\rangle_{\widehat{\rho}} = \int_{\mathbb{R}^{2n}} a(z)\rho(z)dz \tag{14.13}$$

where ρ *is the Wigner distribution of the density operator* $\widehat{\rho}$.

Proof. Formula (14.13) immediately follows from the product formula (12.16) since $\widehat{\rho} = (2\pi\hbar)^n\,\mathrm{Op_W}(\rho)$. □

Using the decomposition

$$\rho = \sum_j \lambda_j W\psi_j$$

in a direct sum, we see that

$$\langle\widehat{A}\rangle_{\widehat{\rho}} = \sum_j \lambda_j \int_{\mathbb{R}^{2n}} a(z)W\psi_j(z)dz = \sum_j \lambda_j\langle\widehat{A}\rangle_{\psi_j}. \tag{14.14}$$

14.4 Quantum condition on the covariance matrix

14.4.1 The covariance matrix of a density operator

Let ρ be a real valued-function defined on $\mathbb{R}^{2n}$ satisfying the normalization condition

$$\int_{\mathbb{R}^{2n}} \rho(z)dz = 1 \tag{14.15}$$

(ρ is not necessarily the Wigner distribution of a density matrix $\widehat{\rho}$ at this point). We will call ρ a quasi-distribution on $\mathbb{R}^{2n}$. We assume in addition that ρ decreases sufficiently fast at infinity to guarantee that

$$\int_{\mathbb{R}^{2n}} x^{\alpha} p^{\beta} |\rho(x,p)| dpdx < \infty \quad \text{if } |\alpha| + |\beta| \leq 2. \tag{14.16}$$

In particular, we have $\rho \in L^1(\mathbb{R}^{2n})$.

Let us write $z = (x,p)$ as $z = (z_1, \ldots, z_{2n})$, that is $z_\alpha = x_\alpha$ if $1 \leq \alpha \leq n$ and $z_\alpha = p_\alpha$ if $n+1 \leq \alpha \leq 2n$.

Definition 182. Let ρ be defined as just stated. The covariance matrix associated with ρ is the real symmetric $2n \times 2n$ matrix defined by

$$\Sigma = \int_{\mathbb{R}^{2n}} (z - \langle z \rangle)(z - \langle z \rangle)^T \rho(z)dz \tag{14.17}$$

where $\langle z \rangle$, the average (or expectation) vector, is given by

$$\langle z \rangle = \int_{\mathbb{R}^{2n}} z\rho(z)dz. \tag{14.18}$$

(All vectors z are viewed as column matrices in this definition).

There is a multitude of objects associate with the covariance matrix; they are defined following the lines of classical statistics. For instance,

$$\langle z_\alpha \rangle_\rho = \int_{\mathbb{R}^{2n}} z_\alpha \rho(z)dz$$

is called the α-th expectation of ρ. The number

$$\Sigma_{\alpha\beta} = \int_{\mathbb{R}^{2n}} (z_\alpha - \langle z_\alpha \rangle_\rho)(z_\beta - \langle z_\beta \rangle_\rho)\rho(z)dz \tag{14.19}$$

is called the (α, β) covariance of ρ, so that

$$\Sigma = (\Sigma_{\alpha\beta})_{1 \leq \alpha,\beta \leq 2n}.$$

When $\alpha = \beta$, the number $\Sigma_\alpha^2 = \Sigma_{\alpha\alpha}$ is called the α-th variance of ρ.

It will often be convenient to write the covariance matrix in block form as

$$\Sigma = \begin{pmatrix} \Sigma_{XX} & \Sigma_{XP} \\ \Sigma_{PX} & \Sigma_{PP} \end{pmatrix}, \quad \Sigma_{PX} = \Sigma_{XP}^T$$

where the $n \times n$ blocks $\Sigma_{XX}, \Sigma_{XP}, \Sigma_{PP}$ are given by $\Sigma_{XX} = (\Sigma_{\alpha\beta})_{1\leq\alpha,\beta\leq n}$, etc.

14.4.2 The quantum condition

We recall (Theorem 185) that a necessary condition for an operator trace class $\widehat{\rho}$ on $L^2(\mathbb{R}^n)$ to be positive semi-definite (and hence a density operator) is that its covariance matrix Σ satisfies (when defined) the quantum condition

$$\Sigma + \frac{i\hbar}{2} J \geq 0. \tag{14.20}$$

Also recall that this condition is equivalent to

$$\hbar \leq 2\lambda_{\min} \tag{14.21}$$

where $\lambda_{\min}$ is the smallest symplectic eigenvalue of Σ.

We now address the sensitivity of this condition on the parameter $\hbar$. We begin by noticing the following rather straightforward result:

Proposition 183. *If the quantum condition (14.20) holds, then it holds for every $\eta < \hbar$:*

$$\Sigma + \frac{i\hbar}{2} J \geq 0 \Longrightarrow \Sigma + \frac{i\eta}{2} J \geq 0 \textit{ for } \eta < \hbar.$$

Proof. Setting $\eta = r\hbar$ with $0 < r \leq 1$, we have

$$\Sigma + \frac{i\eta}{2} J = (1 - r)\Sigma + r\left(\Sigma + \frac{i\hbar}{2} J\right) \geq 0$$

as soon as $r \leq 1$ since the sum of two ≥ 0 matrices is also ≥ 0. □

Note that this property does not allows us to conclude that a quantum state remains a quantum state if we decrease $\hbar$ because the quantum condition (14.20) is necessary but not sufficient to insure we have a *bona fide* density operator (except for Gaussian states).

We are going to prove a very important result connecting the covariance matrix formalism with the positivity issues studied in Chapter 13. The proof relies on the following technical lemma:

Lemma 184. *If the function* $a : \mathbb{R}^{2n} \longrightarrow \mathbb{C}$ *is continuous and twice continuously differentiable near* 0 *and of* $\hbar$*-positive type, then we have*

$$-2\hbar a''(0) + iJ \geq 0 \tag{14.22}$$

where $a''(0) = D^2 a(0)$ *is the Hessian matrix of* a *at* 0.

Proof. For $(\lambda_1, \ldots, \lambda_m) \in \mathbb{C}^m$ and $\varepsilon \in \mathbb{R}$, let us define

$$R(\varepsilon) = \sum_{j,k=1}^{m} \overline{\lambda_j} \lambda_k e^{-\frac{i\varepsilon^2}{2\hbar}\sigma(z_j,z_k)} a(\varepsilon(z_j - z_k)).$$

If a is of $\hbar$-positive type, we have $R(\varepsilon) \geq 0$ for every ε; choose now the λ_j such that $\sum_j \lambda_j = 0$; then $R(0) = 0$ and $R''(0) \geq 0$. An elementary calculation shows that

$$R''(0) = u^T(-2a''(0) + i\hbar^{-1}J)u$$

where $u = \sum_j \lambda_j z_j \in \mathbb{C}^{2n}$. The λ_j and z_j being arbitrary we thus have $-2a''(0) + i\hbar^{-1}J \geq 0$, proving the inequality (14.22). □

Let us now prove:

Theorem 185 (Quantum condition). *Let* ρ *be the Wigner distribution of a Feichtinger state* $\widehat{\rho}$. (i) *The covariance matrix* Σ *satisfies the "quantum condition"*

$$\Sigma + \frac{i\hbar}{2}J \geq 0 \tag{14.23}$$

i. e., the Hermitian matrix $\Sigma + \frac{i\hbar}{2}J$ *is positive semidefinite;* (ii) *the matrix* Σ *is positive:* $\Sigma > 0$ *(and hence invertible);* (iii) *condition (14.23) is also sufficient for* $\widehat{\rho}$ *to be a Feichtinger state when* ρ *is a Gaussian*

$$\rho(z) = (2\pi)^{-n}\sqrt{\det \Sigma^{-1}} e^{-\frac{1}{2}\Sigma^{-1}z^2}. \tag{14.24}$$

Proof. (i) That $\Sigma + \frac{i\hbar}{2}J$ is Hermitian is clear: Since the adjoint of J is $-J$, we have

$$\left(\Sigma + \frac{i\hbar}{2}J\right)^* = \Sigma^* + \left(\frac{i\hbar}{2}J\right)^* = \Sigma + \frac{i\hbar}{2}J.$$

We next remark that $\Sigma = \Sigma_0$ where Σ_0 is the covariance matrix of $\rho_0(z) = \rho(z + \langle z \rangle_\rho)$: We have $\langle z \rangle_{\rho_0} = 0$, and hence it is thus sufficient to assume that the density operator $\widehat{\rho}$ is centered at 0. We are going to use Lemma 184 with $f(z) = F_\sigma \rho_0(z) = \rho_{0,\sigma}(z)$. Observing that the symplectic Fourier transform ρ_σ is twice continuously differentiable in view of the argument preceding the statement of the theorem, we have

$$\hbar^2 \rho_\sigma''(0) = (2\pi\hbar)^{-n} \begin{pmatrix} -\Sigma_{PP} & \Sigma_{XP} \\ \Sigma_{PX} & -\Sigma_{XX} \end{pmatrix},$$

and hence

$$\hbar^2 \rho''_\sigma(0) = \left(\frac{1}{2\pi\hbar}\right)^n J\Sigma J. \tag{14.25}$$

Since $\widehat{\rho}$ is a density matrix, we have

$$M = -2\hbar^{-1} J\Sigma J + iJ \geq 0;$$

the condition $M \geq 0$ being equivalent to $J^T MJ \geq 0$ the inequality (14.23) follows. Let us show that $\Sigma > 0$. (ii) Suppose *in contrario* that Σ has a negative eigenvalue λ, and let z_λ be a real eigenvector for λ. Since $Jz_\lambda \cdot z_\lambda = 0$, we have

$$\left(\Sigma + \frac{i\hbar}{2} J\right) z_\lambda \cdot z_\lambda = \Sigma z_\lambda \cdot z_\lambda = \lambda |z_\lambda|^2 < 0$$

which contradicts the assumption $\Sigma + \frac{i\hbar}{2} J \geq 0$. There remains to show that 0 cannot be an eigenvalue of Σ. Suppose that 0 is an eigenvalue, and let z_0 be a corresponding real eigenvector. For $\varepsilon > 0$, set $z(\varepsilon) = (I + i\varepsilon J)z_0$. Using the relationships $\Sigma z_0 = 0$ and $Jz_0 \cdot z_0 = \sigma(z_0, z_0) = 0$, we get, after a few calculations,

$$\left(\Sigma + \frac{i\hbar}{2} J\right) z(\varepsilon) \cdot z(\varepsilon) = \varepsilon \frac{1}{2} \hbar |z_0|^2 + \varepsilon^2 \Sigma (Jz_0) \cdot Jz_0.$$

Choose now ε opposite in sign to $\frac{1}{2}\hbar$; then $\varepsilon \frac{1}{2}\hbar |z_0|^2 < 0$ and, if $|\varepsilon|$ is small enough, we have $z(\varepsilon)^T (\Sigma + \frac{i\hbar}{2} J) z(\varepsilon) < 0$, which implies that $\Sigma + \frac{i\hbar}{2} J > 0$. (iii) The sufficiency of the quantum condition is an immediate consequence of Proposition 168 since the $\hbar$-positivity of ρ implies that $\widehat{\rho} \geq 0$, hence $\widehat{\rho}$ is a density matrix. □

Remark 186. We will show later (Proposition 200) that the quantum condition $\Sigma + \frac{i\hbar}{2} J \geq 0$ has a geometric reformulation in terms of the symplectic eigenvalues of the covariance matrix.

We stress that the condition (14.23) is *necessary* but *not sufficient* for an arbitrary phase-space function to be the Wigner distribution of a quantum state (except when ρ is a Gaussian (14.24)). Let us check this on the following counterexample. Consider the function $f(x, p)$ defined for $n = 1$ by

$$f(x, p) = \left(1 - \frac{1}{2} a x^2 - \frac{1}{2} b x^2\right) e^{-(a^2 x^4 - b^2 p^4)}$$

where a and b are positive constants such that $ab \geq \frac{1}{4}\hbar^2$. Define now

$$\rho(x, p) = \frac{1}{2\pi} \int_{\mathbb{R}^2} e^{-i(xx' + pp')} f(x', p') dp' dx';$$

since f is an even function, ρ is real. In view of the Fourier inversion formula, we have

$$f(x,p) = \int_{\mathbb{R}^2} e^{i(xx'+pp')}\rho(x',p')dp'dx',$$

and hence

$$\int_{\mathbb{R}^2} \rho(x,p)dpdx = f(0,0) = 1$$

so that $\rho(x,p)$ is a candidate for being the Wigner function of some density matrix. Calculating the covariance matrix Σ associated with ρ, one finds after some tedious but straightforward calculations that $\Sigma + \frac{i\hbar}{2}J \geq 0$. However, ρ cannot be the Wigner function of a density matrix $\widehat{\rho}$ because, if this were the case, we would have $\langle p^4\rangle_{\widehat{\rho}} \geq 0$; but by definition of ρ,

$$\langle p^4\rangle_{\widehat{\rho}} = \int_{\mathbb{R}^2} p^4\rho(x,p)dpdx = \frac{\partial^4 f}{\partial x^4}(0,0) = -24a^2 < 0.$$

14.5 The Narcowich–Wigner spectrum

Let $\rho \in L^2(\mathbb{R}^{2n})$; we assume that ρ is real and that

$$\int_{\mathbb{R}^{2n}} \rho(z)dz = 1. \tag{14.26}$$

We also assume that the symplectic η-Fourier transform

$$\rho_{\sigma,\eta}(z) = \left(\frac{1}{2\pi\eta}\right)^n \int_{\mathbb{R}^{2n}} e^{-\frac{i}{\eta}\sigma(z,z')}\rho(z')dz'$$

($\eta \in \mathbb{R}$, $\eta \neq 0$) is continuous.

Definition 187. The Narcowich–Wigner spectrum $\Sigma_{\mathrm{NW}}(\rho)$ consists of all $\eta \in \mathbb{R}$, $\eta \neq 0$ such that ρ is the η-Wigner distribution of a density operator $\widehat{\rho}_\eta$.

The Narcowich–Wigner has the following elementary properties (where one allows the parameter η to take negative values):

Proposition 188. *Let ρ be as just stated.* (i) *We have $\eta \in \Sigma_{\mathrm{NW}}(\rho)$ if and only if $-\eta \in \Sigma_{\mathrm{NW}}(\rho)$;* (ii) *the Narcowich–Wigner spectrum of ρ is a bounded subset of $\mathbb{R}$: There exists $\eta_0 \in \mathbb{R}$ such that $\Sigma_{\mathrm{NW}}(\rho) \subset [-\eta_0 0) \cup (0, \eta_0]$.*

The following result was proved by Dias and Prata:

Proposition 189. *Assume that ρ is not a Gaussian distribution. Then* $\Sigma_{NW}(\rho) = \{-\hbar, \hbar\}$.

For a proof of this result, see the comments and references section; the last result is intriguing because one can wonder why, as soon as a state becomes non-Gaussian, its Narcowich–Wigner spectrum collapses in such a spectacular way. A clue might be the extremality properties of Gaussian states that singles them out among all other states.

14.6 Comments and references

The quantum condition (Theorem 185) has a long and rather chaotic story. The first to use the positivity condition $\Sigma + \frac{i\hbar}{2}J \geq 0$ on the covariance matrix were probably Arvind, Dutta, Mukunda, and Simon in [4]. They did that in order to reformulate the uncertainty principle (which we study in Chapter 15). It has become one of the main tools in quantum optics and quantum information theory because it serves as a test replacing the Robertson–Schrödinger inequalities, which are difficult to handle in practice. The main contributions to the proof of Theorem 185 are due to Narcowich and his collaborators [58–60].

The Narcowich–Wigner spectrum $\Sigma_{NW}(\rho)$ was introduced by Narcowich in [58] and further studied by Bröcker and Werner [13]. Mathematically speaking the topic of variable values for $\hbar$ is very much open, and very little is known about the Narcowich–Wigner spectrum.

15 The uncertainty principle

We pursue the study of covariance matrix of a density matrix and examine its relationship with the uncertainty principle of quantum mechanics, and show that it can be expressed in terms of a deep property of symplectic topology. We also suggest a new more general geometric version of the uncertainty principle that involves the use of polarity between convex sets and appears naturally when one studies the orthogonal projections of the covariance ellipsoid on the x and p subspaces.

15.1 Feichtinger states

Let us introduce the following terminology:

Definition 190. A mixed quantum state $\{(\psi_j, \lambda_j)\}$ in $L^2(\mathbb{R}^n)$ is called a Feichtinger (quantum) state of order s if each ψ_j belongs to the modulation space $M_s^1(\mathbb{R}^n)$. Identifying the state $\{(\psi_j, \lambda_j)\}$ with the density operator

$$\widehat{\rho} = \sum_j \lambda_j \widehat{\rho}_{\psi_j},$$

we will also call $\widehat{\rho}$ a Feichtinger state of order s.

One needs to be somewhat careful at this point. Each mixed quantum state $\{(\psi_j, \lambda_j)\}$ defines a unique density operator $\widehat{\rho}$, but the converse of this property is not true because mixed quantum states can usually be decomposed in infinitely many ways. For instance, the spectral theorem tells us that $\widehat{\rho}$ can be written

$$\widehat{\rho} = \sum_j \lambda'_j \widehat{\rho}_{\psi'_j}$$

where the ψ'_j form an orthonormal system; now, the definition of a Feichtinger state does not imply that $\psi'_j \in M_s^1(\mathbb{R}^n)$ if $\psi_j \in M_s^1(\mathbb{R}^n)$. The definition thus makes use of one particular decomposition of the density operator.

Recall that the condition $\psi \in M_2^1(\mathbb{R}^n)$ means that $\psi \in L^2(\mathbb{R}^n)$ and

$$\int_{\mathbb{R}^{2n}} |W\psi(z)(1 + |z|^2) dz < \infty. \tag{15.1}$$

It follows that the Wigner distribution

$$\rho(z) = \sum_j \lambda_j W\psi_j(z)$$

of a Feichtinger state satisfies

$$\int_{\mathbb{R}^{2n}} (1 + |z|^2)|\rho(z)| dz < \infty. \tag{15.2}$$

https://doi.org/10.1515/9783110722772-015

An immediate consequence is that the conditions (14.16) ensuring the existence of Σ are satisfied, hence the covariance matrix of a Feichtinger state is always well-defined.

Here is an interesting differentiability property satisfied by the Wigner distributions of Feichtinger that states:

Proposition 191. *The Fourier transforms $F\rho$ and $\rho_\sigma = F_\sigma\rho$ of the Wigner distribution of a Feichtinger state with $s \geq 2$ are twice continuously differentiable: $F\rho \in C^2(\mathbb{R}^{2n})$.*

Proof. It suffices to assume that $s = 2$. Since $\rho_\sigma(z) = F\rho(Jz)$, it suffices to prove the result for the ordinary Fourier transform F, that is

$$F\rho(z) = \left(\frac{1}{2\pi\hbar}\right)^n \int_{\mathbb{R}^{2n}} e^{-\frac{i}{\hbar}z\cdot z'}\rho(z')dz'.$$

Since (15.2) is satisfied, we may differentiate twice under the integration sign, and we get

$$\partial_{z_\alpha} F\rho = -\frac{i}{\hbar}F(z_\alpha\rho), \quad \partial_{z_\alpha}\partial_{z_\beta} F\rho = \left(-\frac{i}{\hbar}\right)^2 F(z_\alpha z_\beta\rho),$$

and hence

$$|\partial_{z_\alpha} F\rho(z)| \leq \frac{1}{\hbar}\left|\int_{\mathbb{R}^{2n}} z_\alpha\rho(z)dz\right| < \infty$$

$$|\partial_{z_\alpha}\partial_{z_\beta} F\rho(z)| \leq \left(\frac{1}{\hbar}\right)^2\left|\int_{\mathbb{R}^{2n}} z_\alpha z_\beta\rho(z)dz\right| < \infty.$$ □

15.2 The Robertson–Schrödinger inequalities

The Robertson–Schrödinger inequalities are closely related to the quantum condition in Theorem 185. This connection is after all hardly surprising, as shown by the following result:

Proposition 192. *Let Σ be a real symmetric $2n \times 2n$ matrix. Suppose that Σ satisfies the quantum condition*

$$\Sigma + \frac{i\hbar}{2}J \geq 0. \tag{15.3}$$

Then the inequalities

$$(\Delta X_j)^2(\Delta P_j)^2 \geq \Delta(X_j, P_j)^2 + \frac{1}{4}\hbar^2 \tag{15.4}$$

hold for $1 \leq j \leq n$.

Proof. Recall that the matrix $\Sigma + \frac{i\hbar}{2}J$ is Hermitian and that $\Sigma > 0$. The non-negativity of $\Sigma + \frac{i\hbar}{2}J$ can be expressed in terms of the minors

$$\Sigma_j = \begin{pmatrix} (\Delta X_j)^2 & \Delta(X_j, P_j) + \frac{i\hbar}{2} \\ \Delta(P_j, X_j) - \frac{i\hbar}{2} & (\Delta P_j)^2 \end{pmatrix}$$

which are non-negative provided that $\Sigma + \frac{i\hbar}{2}J$ is. Since

$$\operatorname{Tr}(\Sigma_j) = (\Delta X_j)^2 + (\Delta P_j)^2 \geq 0,$$

we have $\Sigma_j \geq 0$ if and only if

$$\det \Sigma_j = (\Delta X_j)^2(\Delta P_j)^2 - \Delta(X_j, P_j)^2 - \frac{1}{4}\hbar^2 \geq 0$$

which is equivalent to the inequality (15.4). □

The inequalities (15.4) are called the *Robertson–Schrödinger inequalities* in the physical literature; notice that they imply the textbook "Heisenberg uncertainty principle"

$$(\Delta X_j)(\Delta P_j) \geq \frac{1}{2}\hbar, \quad 1 \leq j \leq n. \tag{15.5}$$

At this point, it is appropriate to notice that (except in the case $n = 1$) the condition $\Sigma + \frac{i\hbar}{2}J \geq 0$ is *not in general equivalent* to the uncertainty inequalities (15.4); it is in fact a *stronger* condition except for $n = 1$. In this case, the covariance matrix is just

$$\Sigma = \begin{pmatrix} \Delta X^2 & \Delta(X, P) \\ \Delta(P, X) & \Delta P^2 \end{pmatrix},$$

and since

$$\operatorname{Tr}\left(\Sigma + \frac{i\hbar}{2}J\right) = \Delta X^2 + \Delta P^2 \geq 0,$$

the condition $\Sigma + \frac{i\hbar}{2}J \geq 0$ is equivalent to $\det(\Sigma + \frac{i\hbar}{2}J) \geq 0$, that is, to

$$\Delta X^2 \Delta P^2 - \Delta(X, P)^2 \geq \frac{1}{4}\hbar^2$$

which is precisely (15.4) for $n = 1$. It is however important to notice that the equivalence between $\Sigma + \frac{i\hbar}{2}J \geq 0$ and the uncertainty inequalities (15.4) no longer holds for $n > 1$.

The following result is well-known and proven in elementary books using statistical methods and the Cauchy–Schwarz inequality:

Corollary 193. *Let $\widehat{\rho}$ be a Feichtinger state on $L^2(\mathbb{R}^n)$ and Σ its covariance matrix. Then the associated covariances satisfy the Robertson–Schrödinger inequalities*

$$(\Delta X_j)^2(\Delta P_j)^2 \geq \Delta(X_j, P_j)^2 + \frac{1}{4}\hbar^2.$$

Proof. It is an immediate consequence of Theorem 185 and of Proposition 192. □

Notice that the assumption that $\widehat{\rho}$ is a Feichtinger state is essential since it guarantees the existence of the covariance matrix.

15.3 Gromov's symplectic non-squeezing theorem

15.3.1 Statement of Gromov's theorem

We will study in next section the uncertainty principle from a geometric point of view. The key to our main result (Theorem 201) is a very deep property from symplectic topology due to Mikhail Gromov. Before we state his result, let us introduce the following notation: For $R > 0$, we denote by

$$B^{2n}(R) = \{z \in \mathbb{R}^{2n} : |z| \leq R\}$$

the phase space ball with center 0 and radius R, and by

$$Z_j^{2n}(R) = \{z = (x,p) \in \mathbb{R}^{2n} : x_j^2 + p_j^2 \leq R\},$$

the cylinder with radius R centered at 0 and based on the x_j, p_j plane. When centered at an arbitrary point $z_0 \in \mathbb{R}^{2n}$, these sets are denoted by $B^{2n}(z_0, R)$ and $Z_j^{2n}(z_0, R)$. Recall that a symplectomorphism f of $(\mathbb{R}^{2n}, \sigma)$ is a diffeomorphism of $\mathbb{R}^{2n}$ respecting the symplectic structure: $f^*\sigma = \sigma$ (i. e., the Jacobian matrix $Df(z) \in \operatorname{Sp}(n)$ for every $z \in \mathbb{R}^{2n}$).

Theorem 194 (Gromov). *There exists a symplectomorphism f of $(\mathbb{R}^{2n}, \sigma)$ such that $f(B^{2n}(z_0, R)) \subset Z_j^{2n}(z_0, r)$ $(j = 1, \ldots, n)$ if and only if $R \leq r$.*

The choice of index j for the cylinders $Z_j^{2n}(z_0, r)$ is irrelevant because $Z_j^{2n}(z_0, r)$ can always been taken to another cylinder $Z_k^{2n}(z_0, r)$ using a trivial linear symplectic transformation. All the known proofs (direct or indirect) of Gromov's theorem are notoriously difficult; Gromov's original proof makes use of the theory of pseudo-holomorphic curves. We will actually only need its linear variant, which is somewhat easier to prove.

15.3.2 Proof of Gromov's theorem in the linear case

We will need the following elementary lemma from linear algebra:

Lemma 195. *Let $S = \left(\begin{smallmatrix} A & B \\ C & D \end{smallmatrix}\right)$ be a $2n \times 2n$ symplectic matrix, and let $(a, b) = (a_1, \ldots, a_n, b_1, \ldots, b_n)$, $(c, d) = (c_1, \ldots, c_n, d_1, \ldots, d_n)$ its j-th line and $(n+j)$-th line, respectively. We have*

$$a \cdot d - b \cdot c = 1.$$

Proof. Since S is symplectic, A, B, C and D satisfy the conditions

$$AD^T - BC^T = I.$$

(see (1.14)). The equality $a \cdot d - b \cdot c = 1$ follows. □

Let us now prove the linear Gromov theorem:

Proposition 196. *There exists $S \in \mathrm{Sp}(n)$ such that $S(B^{2n}(R)) \subset Z_j^{2n}(r)$ if and only if $R \leq r$.*

Proof. Since S is linear, it is sufficient, by homogeneity, to assume that $R = 1$. It is moreover no restriction to assume $j = 1$. It is thus sufficient to prove the implication

$$\left.\begin{array}{c} x_1^2 + p_1^2 \leq r^2 \\ (x,p) \in S(B^{2n}(1)) \end{array}\right\} \Rightarrow 1 \leq r. \tag{15.6}$$

Denoting by a, b, c, d the first lines of the matrices A, B, C, D, respectively, we have

$$x_1 = a \cdot x + b \cdot p, \quad p_1 = c \cdot x + d \cdot p.$$

Setting $u = (a, b)$ and $v = (c, d)$, the condition $x_1^2 + p_1^2 < r^2$ for all $(x, p) \in S(B^{2n}(1))$ is thus equivalent to

$$(u \cdot z)^2 + (v \cdot z)^2 \leq r^2$$

for all $z = (x, p)$ with $|z| \leq 1$. In particular, choosing respectively $z = u/|u|$ and $z = v/|v|$, we must thus have both $|u|^2 \leq r^2$ and $|v|^2 \leq r^2$, that is

$$a \cdot a + b \cdot b \leq r^2 \quad \text{and} \quad c \cdot c + d \cdot d \leq r^2. \tag{15.7}$$

Let us show that these inequalities imply that we must have $1 \leq r^2$; the result will follow. In view of Lemma 195, it follows by Cauchy–Schwarz's inequality that

$$1 = a \cdot d - b \cdot c \leq (a^2 + b^2)^{1/2}(c^2 + d^2)^{1/2} \leq r^2, \tag{15.8}$$

which implies that $r \geq 1$. □

15.3.3 The Gromov width

Associated with Gromov's symplectic non-squeezing theorem comes the notion of Gromov width. We again limit ourselves to the linear case.

Definition 197. The (linear) Gromov width of a subset $\Omega \subset \mathbb{R}^{2n}$ is the (possibly infinite) number

$$w_G(\Omega) = \sup_{S \in \mathrm{Sp}(n), z_0 \in \mathbb{R}^{2n}} \{\pi R^2 : S(B^{2n}(z_0, R)) \subset \Omega\}.$$

It is obvious that $w_G(B^{2n}(z_0,R)) = \pi R^2$. It turns out that it follows from Gromov's (linear) theorem just proven that we actually also have $w_G(Z_j^{2n}(z_0,R)) = \pi R^2$. In view of the obvious implication

$$\Omega' \subset \Omega \Longrightarrow w_G(\Omega') \leq w_G(\Omega),$$

we thus have

$$B^{2n}(z_0,R) \subset \Omega \subset Z_j^{2n}(z_0,R) \Longrightarrow w_G(\Omega) = \pi R^2.$$

The Gromov width is a particular case of the more general notion of symplectic capacity (see the comments and references section).

15.4 The covariance ellipsoid

The notion of covariance ellipsoid will allow us to give geometric interpretations of the uncertainty principle. Let us introduce the following terminology:

Definition 198. Let Σ be a quantum covariance matrix (hence $\Sigma > 0$). The phase space ellipsoid

$$\Omega = \left\{ z \in \mathbb{R}^{2n} : \frac{1}{2}\Sigma^{-1}z^2 \leq 1 \right\}$$

is called the covariance ellipsoid associated with Σ.

One also sometimes uses the term "Wigner ellipsoid" for Ω.

We are going to prove a very important reformulation of the uncertainty principle geometric in terms of concepts from symplectic geometry. Let us first give the following definition:

Definition 199. (i) Let $B^{2n}(R)$ be the closed phase space ball with radius R centered at 0. The image $S(B^{2n}(R))$ by $S \in \mathrm{Sp}(n)$ is called a symplectic ball. (ii) When $R = \sqrt{\hbar}$, the symplectic ball $S(B^{2n}(\sqrt{\hbar}))$ is called a "quantum blob".

Here is a very useful reformulation of the quantum condition $\Sigma + \frac{i\hbar}{2}J \geq 0$ in terms of the symplectic eigenvalues of the covariance matrix:

Proposition 200. *The quantum condition $\Sigma + \frac{i\hbar}{2}J \geq 0$ is equivalent to the inequality*

$$\hbar \leq 2\lambda_{\min} \tag{15.9}$$

where $\lambda_{\min}$ is the smallest symplectic eigenvalue of Σ.

Proof. Let us show that the conditions (15.9) and (15.3) indeed are equivalent. Let $\Sigma = S^T DS$ be a Williamson diagonalization of Σ, that is $S \in \mathrm{Sp}(n)$ and $D = \begin{pmatrix} \Lambda & 0 \\ 0 & \Lambda \end{pmatrix}$ where Λ is the diagonal matrix whose diagonal entries are the symplectic eigenvalues $\lambda_1, \ldots, \lambda_n$ of Σ. Since $S^T JS = J$, condition (15.3) is equivalent to $D + \frac{i\hbar}{2}J \geq 0$; the characteristic polynomial of $D + \frac{i\hbar}{2}J$ is the determinant

$$\begin{aligned} P(\lambda) &= \begin{vmatrix} \Lambda - \lambda I_n & \frac{i\hbar}{2}I_n \\ -\frac{i\hbar}{2}I_n & \Lambda - \lambda I_n \end{vmatrix} \\ &= \det\left[(\Lambda - \lambda I_n)^2 - \frac{1}{4}\hbar^2 I_n \right]; \end{aligned}$$

the matrix Λ being diagonal, the zeroes λ of $P(\lambda)$ are the solutions of the n equations $(\lambda_j - \lambda)^2 - \frac{1}{4}\hbar^2 = 0$, that is $\lambda_j - \lambda = \pm\frac{1}{2}|\hbar|$. Since $\lambda \geq 0$, we must have $\lambda_j \geq \frac{1}{2}|\hbar|$ for all j, hence (15.9). □

We now have the material for proving our geometric reformulation of the uncertainty principle. We give it is the status of a theorem because of its great usefulness.

Theorem 201. (i) *A covariance matrix* Σ *satisfies the quantum condition* $\Sigma + \frac{i\hbar}{2}J \geq 0$ *if and only if the associated covariance ellipsoid* Ω *contains a quantum blob.* (ii) *This condition is equivalent to the inequality* $w_G(\Omega) \geq \pi\hbar$.

Proof. (i) As in the proof of Lemma 200, a Williamson diagonalization $\Sigma = SDS^T$ (observe that S^T is written before S) brings the quantum condition in the form $D + \frac{i\hbar}{2}J \geq 0$ where

$$D = \begin{pmatrix} \Lambda & 0 \\ 0 & \Lambda \end{pmatrix}, \quad \Lambda = \operatorname{diag}(\lambda_1, \ldots, \lambda_n),$$

the λ_j being the symplectic eigenvalues of Σ. The condition $\frac{1}{2}\Sigma^{-1}z \cdot z \leq 1$ is equivalent to

$$\frac{1}{2}(S^T)^{-1}D^{-1}S^{-1}z \cdot z \leq 1$$

so that the $\Omega = S(\Omega_D)$ where Ω_D is the ellipsoid defined by $\frac{1}{2}D^{-1}z^2 \leq 1$; in x_j, p_j coordinates,

$$\Omega_D : \sum_{j=1}^{n} \frac{1}{2\lambda_j}(x_j^2 + p_j^2) \leq 1.$$

Assume now that $S'(B^{2n}(\sqrt{\hbar})) \subset \Omega_D$ for some $S' \in \mathrm{Sp}(n)$. In view of Gromov's non-squeezing theorem, the areas of the orthogonal projections of $S'(B^{2n}(R))$ on every x_j, p_j plane must be at least $\pi\hbar$, but the condition $S'(B^{2n}(\sqrt{\hbar})) \subset \Omega_D$ is equivalent to saying the sections $x_j^2 + p_j^2 \leq 2\lambda_j$ of Ω_D by these planes must have are $2\pi\lambda_j \geq \pi\hbar$, for all $j = 1, \ldots, n$. This is equivalent to $\hbar \leq 2\lambda_{\min}$, hence the result in view of the inequality (15.9) in Proposition 200 since the condition $S'(B^{2n}(\sqrt{\hbar})) \subset \Omega_D$ means that $SS'(B^{2n}(\sqrt{\hbar})) \subset$

$S(\Omega_D) = \Omega$. (ii) Suppose that $S(B^{2n}(\sqrt{\hbar})) \subset \Omega$ for some $S \in \mathrm{Sp}(n)$; by the monotonicity and symplectic invariance of the Gromov width, we have

$$\pi\hbar = w_G(B^{2n}(\sqrt{\hbar})) = w_G(S(B^{2n}(\sqrt{\hbar}))) \le w_G(\Omega)$$

so that $w_G(\Omega) \ge \pi\hbar$. If, conversely, $w_G(\Omega) \ge \pi\hbar$ then the supremum of all radii R such that $S(B^{2n}(R)) \subset \Omega$ is $\sqrt{\hbar}$, hence $S(B^{2n}(\sqrt{\hbar})) \subset \Omega$. □

15.5 Uncertainty and quantum polarity

15.5.1 Orthogonal projections of the covariance ellipsoid

The projection of an ellipsoid on any subspace is an ellipsoid. We are going to focus here on the orthogonal projections of the covariance ellipsoid on the position and momentum subspaces $\mathbb{R}^n_x \equiv \mathbb{R}^n \times 0$ and $\mathbb{R}^n_p \equiv 0 \times \mathbb{R}^n$ of phase space $\mathbb{R}^{2n}$. We will denote these projections by

$$\Pi_X : \mathbb{R}^{2n} \longrightarrow \mathbb{R}^n_x, \quad \Pi_X(x,p) = x$$
$$\Pi_P : \mathbb{R}^{2n} \longrightarrow \mathbb{R}^n_p, \quad \Pi_P(x,p) = p.$$

We begin with the following general result:

Lemma 202. *Let $\Omega = \{z : Mz \cdot z \le \hbar\}$, $M > 0$. The orthogonal projections $\Omega_X = \Pi_X\Omega$ and $\Omega_P = \Pi_P\Omega$ are the ellipsoids*

$$\Omega_X = \{x \in \mathbb{R}^n_x : (M/M_{PP})x^2 \le \hbar\} \tag{15.10}$$

$$\Omega_P = \{p \in \mathbb{R}^n_p : (M/M_{XX})p^2 \le \hbar\} \tag{15.11}$$

where

$$M/M_{PP} = M_{XX} - M_{XP}M_{PP}^{-1}M_{PX} \tag{15.12}$$

$$M/M_{XX} = M_{PP} - M_{PX}M_{XX}^{-1}M_{XP} \tag{15.13}$$

are the Schur complements in M of M_{PP} and M_{XX}, respectively.

Proof. Let us set $Q(z) = Mz^2 - \hbar$; the boundary $\partial\Omega$ of the hypersurface $Q(z) = 0$ is defined by

$$M_{XX}x^2 + 2M_{PX}x \cdot p + M_{PP}p^2 = \hbar. \tag{15.14}$$

A point x belongs to the boundary $\partial\Omega_X$ of Ω_X if and only if the normal vector to $\partial\Omega$ at the point $z = (x,p)$ is parallel to $\mathbb{R}^n_x \times 0$, hence we get the constraint $\nabla_z Q(z) = 2Mz \in \mathbb{R}^n_x \times 0$; this is equivalent to saying that $M_{PX}x + M_{PP}p = 0$, that is, to $p = -M_{PP}^{-1}M_{PX}x$. Inserting this value of p in the equation (15.14) shows that $\partial\Omega_X$ is the set of all x such that $(M/M_{PP})x^2 = \hbar$, which yields (15.10). Formula (15.11) is proven in the same way, swapping the subscripts X and P. □

15.5.2 Polar duality in convex geometry

Let X be a convex body in configuration space $\mathbb{R}^n_x$ (a convex body in an Euclidean space is a compact convex set with non-empty interior). We assume in addition that X contains 0 in its interior. This is the case if, for instance, X is symmetric: $X = -X$.

Definition 203. The (quantum) *polar dual* of X is the subset

$$X^\hbar = \{p \in \mathbb{R}^n_p : px \leq \hbar \text{ for all } x \in X\} \tag{15.15}$$

of the dual space $\mathbb{R}^n_p \equiv (\mathbb{R}^n_x)^*$.

Notice that it trivially follows from the definition that $X^\hbar$ is convex. In the mathematical literature, one usually chooses $\hbar = 1$, in which case one writes X^o for the polar dual; we have $X^\hbar = \hbar X^o$. Here is an intuitive interpretation of the polar dual: X being convex, it is the intersection of a (possibly infinite) family of half spaces (the "supporting hyperplanes" of X). Therefore, the polar of X can be seen as the convex hull of a (possibly infinite) set of points, coming from all of the supporting hyperplanes.

The following properties of the polar dual are obvious:

$$\text{Biduality: } (X^\hbar)^\hbar = X; \tag{15.16}$$

$$\text{Antimonotonicity: } X \subset Y \Longrightarrow Y^\hbar \subset X^\hbar; \tag{15.17}$$

$$\text{Scaling: } \det L \neq 0 \Longrightarrow (LX)^\hbar = (L^T)^{-1} X^\hbar. \tag{15.18}$$

The "smaller" X is, the larger $X^\hbar$ is. For instance, if $X = 0$ (corresponding to a perfectly localized system), then $X^\hbar = \mathbb{R}^n_p$, the whole momentum space. This property, reminiscent of the uncertainty principle and of the duality of the support of a function and that of its Fourier transform, becomes particularly visible when one studies the polar duals of ellipsoids. Here are a few useful results:

Proposition 204. *Let $\mathcal{B}^n_X(R)$ (resp. $\mathcal{B}^n_P(R)$) be the ball $\{x : |x| \leq R\}$ in $\mathbb{R}^n_x$ (resp. $\{p : |p| \leq R\}$ in $\mathbb{R}^n_p$).* (i) *We have*

$$\mathcal{B}^n_X(R)^\hbar = \mathcal{B}^n_P(\hbar/R). \tag{15.19}$$

In particular,

$$\mathcal{B}^n_X(\sqrt{\hbar})^\hbar = \mathcal{B}^n_P(\sqrt{\hbar}). \tag{15.20}$$

(ii) *Let $A = A^T$ be an invertible $n \times \times n$ matrix. We have*

$$\{x : Ax^2 \leq R^2\}^\hbar = \{p : A^{-1}p^2 \leq (\hbar/R)^2\}, \tag{15.21}$$

and hence

$$\{x : Ax^2 \leq \hbar\}^\hbar = \{p : A^{-1}p^2 \leq \hbar\}. \tag{15.22}$$

Proof. Let us show that $\mathcal{B}_X^n(R)^\hbar \subset \mathcal{B}_P^n(\hbar/R)$. Let $p \in \mathcal{B}_X^n(R)^\hbar$ and set $x = (R/|p|)p$; we have $|x| = R$, and hence $px \leq \hbar$, that is, $R|p| \leq \hbar$ and $p \in \mathcal{B}_P^n(\hbar/R)$. To prove the opposite inclusion, choose $p \in \mathcal{B}_P^n(\hbar/R)$. We have $|p| \leq \hbar/R$, and hence, by the Cauchy–Schwarz inequality, $px \leq |x||p| \leq \hbar|x|/R$, that is, $px \leq \hbar$ for all x such that $|x| \leq R$; this means that $p \in \mathcal{B}_X^n(R)^\hbar$. (ii) The ellipsoid $\{x : Ax^2 \leq R^2\}^\hbar$ is the image of $\mathcal{B}_X^n(R)$ by the automorphism $A^{-1/2}$; in view of formula (15.21), it follows from the scaling property (15.18) and (15.19) that

$$\{x : Ax^2 \leq R^2\}^\hbar = A^{1/2}\mathcal{B}_X^n(R)^\hbar = A^{1/2}\mathcal{B}_P^n(\hbar/R)$$

which is equivalent to (15.21). □

We have assumed that the convex body X contains the origin 0 in its interior. The definitions and results just listed extend without difficulty to the general case by choosing an arbitrary $x_0 \in X$ and replacing X with $X_0 = -x_0 + X$.

15.5.3 The fundamental property of Ω_X and Ω_P

We are going to show that the orthogonal $\Omega_X = \Pi_X\Omega$ and $\Omega_P = \Pi_P\Omega$ of the covariance ellipsoid form a polar dual pair in the sense just stated, provided that the covariance matrix Σ satisfies the quantum condition $\Sigma + \frac{i\hbar}{2}J \geq 0$.

Theorem 205. *Assume that the covariance ellipsoid Ω satisfies the quantization condition $\Sigma + \frac{i\hbar}{2}J \geq 0$ (resp. $M^{-1} + iJ \geq 0$). Then, the orthogonal projections $\Omega_X = \Pi_X\Omega$ and $\Omega_P = \Pi_P\Omega$ of Ω on $\mathbb{R}_x^n$ and $\mathbb{R}_p^n$, respectively, form a dual quantum pair: we have $\Omega_X^\hbar \subset \Omega_P$.*

Proof. It suffices to prove that there exist $Y \subset \Omega_X$ and $Q \subset \Omega_P$ such that $Y^\hbar \subset Q$. For this purpose, we recall that the quantum condition $\Sigma + (i\hbar/2)J \geq 0$ is equivalent to the existence of $S \in \mathrm{Sp}(n)$ such that $S(\mathcal{B}^{2n}(\sqrt{\hbar})) \subset \Omega$; it is therefore sufficient to show that the projections $Y = \Pi_X(S(\mathcal{B}^{2n}(\sqrt{\hbar})))$ and $Q = \Pi_P(S(\mathcal{B}^{2n}(\sqrt{\hbar})))$ form a quantum pair, that is, $Y^\hbar \subset Q$. The ellipsoid $S(\mathcal{B}^{2n}(\sqrt{\hbar}))$ consists of all $z \in \mathbb{R}_z^{2n}$ such that $Rz^2 \leq \hbar$ where $R = (SS^T)^{-1}$. Since R is symmetric and positive definite, we can write it in block-matrix form as

$$R = \begin{pmatrix} A & B \\ B^T & D \end{pmatrix}$$

with $A > 0$, $D > 0$, and the projections Y and Q are given by formulas (15.10) and (15.11), which read here

$$Y = \{x : (R/D)x^2 \leq \hbar\}$$
$$Q = \{p : (R/A)p^2 \leq \hbar\};$$

we have $R/D > 0$ and $R/A > 0$ [77]. In view of formula (15.22), we have

$$Y^{\hbar} = \{p : (R/D)^{-1}p^2 \leq \hbar\},$$

hence the condition $Y^{\hbar} \subset Q$ is equivalent to

$$(R/D)^{-1} \geq R/A. \tag{15.23}$$

Let us prove that this inequality holds. The conditions $R \in \mathrm{Sp}(n)$, $R = R^T$ being equivalent to $RJR = J$, we have

$$AB^T = BA, \quad B^T D = DB \tag{15.24}$$

$$AD - B^2 = I_{n\times n}. \tag{15.25}$$

These relationships imply that the Schur complements R/D and R/A are

$$R/D = (AD - B^2)D^{-1} = D^{-1} \tag{15.26}$$

$$R/A = A^{-1}(AD - B^2) = A^{-1}, \tag{15.27}$$

and hence the inequality (15.23) holds if and only if $D \geq A^{-1}$. This condition is in turn equivalent to $AD \geq I_{n\times n}$. In fact, the inequality $D \geq A^{-1}$ is equivalent to $A^{1/2}DA^{1/2} \geq I_{n\times n}$; now, $A^{1/2}DA^{1/2}$ and AD have the same eigenvalues, hence $AD \geq I_{n\times n}$. If conversely $AD \geq I_{n\times n}$, then $D^{1/2}AD^{1/2} \geq I_{n\times n}$, hence $A \geq D^{-1}$, that is, $D \geq A^{-1}$. Now (15.25) implies that $AD = I_{n\times n} + B^2$, hence we will have $AD \geq I_{n\times n}$ if $B^2 \geq 0$. To prove that $B^2 \geq 0$ we note that since $AB^T = BA$ (first formula (15.24)), we have $B^T = A^{-1}BA$ so that B and $B^T = B^*$ have the same eigenvalues, and these must be real. It follows that the eigenvalues of B^2 are ≥ 0 hence $B^2 \geq 0$ as claimed. □

15.6 Comments and references

The definition of the notion of Feichtinger state is new and due to the author. That the quantum condition $\Sigma + \frac{i\hbar}{2}J \geq 0$ is not equivalent to the Robertson–Schrödinger inequalities (15.4) is easily seen on counterexamples; see [31], § 6.1.1. This fact can be stated in a lapidary way by saying that "the uncertainty principle does not characterize the quantumness of a state"; see de Gosson and Luef [36, 37] and the references therein.

The notion of polarity is very well-known from convex geometry; see for instance the review paper by [5]; for a detailed account of the relation between quantum uncertainty and polar duality, see our paper [35].

16 Separability and entanglement

A topic of great interest and importance in quantum mechanics is that of separability and of its antinomy, entanglement. Separability and entanglement is the key to many applications (e. g., quantum optics and computing, quantum information, to name a few).

16.1 The reduced density operator

16.1.1 Physical motivation

Suppose we have some information on a mixed state $\{(\psi_j, \lambda_j)\}$ that allows us to declare that, although the state is not known with precision, it must be, with probability λ_j, in a tensor product state $\psi_j^A \otimes \psi_j^B$ where ψ_j^A is defined on $\mathbb{R}^{n_A}$ and ψ_j^B on $\mathbb{R}^{n_B}$ where n_A and n_B are non-zero integers such that $n_A + n_B = n$. The state $\{(\psi_j^A \otimes \psi_j^B, \lambda_j)\}$ is then said to be *separable*; if it is not the case, i. e., if at least one of the pure states ψ_j cannot be written as a tensor product $\psi_j^A \otimes \psi_j^B$, then the state is said to be *entangled*. To decide whether a given mixed state is separable or not is a very difficult question that has not yet been fully resolved at the time of writing. Intimately associated with the notion of separability is that of *reduced density operator*. Imagine a quantum system S_A with Hilbert space H_A (for instance $L^2(\mathbb{R}^{n_A})$) and density operator $\widehat{\rho}^A$. At some initial time (say $t_0 = 0$), the system S_A is coupled to another quantum system S_B with Hilbert space H_B represented by a density operator $\widehat{\rho}^B$ on $\mathcal{H}_B$ so that the total density operator is now $\widehat{\rho}^A \otimes \widehat{\rho}^B$. Assuming that the time evolution of the coupled system is governed by a unitary group $(\widehat{U}_t)$ of transformations of the Hilbert space $H = H_A \otimes H_B$, the density operator of the total system will be $\widehat{U}_t^*(\widehat{\rho}^A \otimes \widehat{\rho}^B)\widehat{U}_t$ at time t. Let us now focus on the original system S_A, thus ignoring the coupled system $S_A \cup S_B$. If there had been no interaction between both systems S_A and S_B in the course of time, the unitary operators $\widehat{U}_t$ would be tensor products $\widehat{U}_t^A \otimes \widehat{U}_t^B$ with $\widehat{U}_t^A$ and $\widehat{U}_t^B$ acting on H_A and H_B, respectively, so the system S_A would be in the state $(\widehat{U}_t^A)^* \widehat{\rho}^A \widehat{U}_t^A$. However, in the general case, we cannot decouple both systems since they are correlated and remain so in time. So, we must in some sense "project" or "restrict" the density operator $\widehat{U}_t^*(\widehat{\rho}^A \otimes \widehat{\rho}^B)\widehat{U}_t$ so that it acts on the Hilbert H_A only. This operation is called "partial tracing" in quantum mechanics, and the density operator on H_A obtained by this procedure is denoted by

$$\widehat{\rho}^A(t) = \mathrm{Tr}_B(\widehat{U}_t^*(\widehat{\rho}^A \otimes \widehat{\rho}^B)\widehat{U}_t).$$

16.1.2 Traditional Hilbert space approach

Let us briefly describe the approach to partial tracing common in books in quantum mechanics. It has the advantage of working in the general setting of arbitrary Hilbert

https://doi.org/10.1515/9783110722772-016

spaces. In what follows, we consider two separable Hilbert spaces $(H_A, (\cdot|\cdot)_A)$ and $(H_B, (\cdot|\cdot)_B)$ and denote their tensor product by $(H, (\cdot|\cdot))$: $H = H_A \otimes H_B$.

Definition 206. Let $\widehat{\rho} \in \mathcal{L}_1(H)$ be a density operator. The partial trace of $\widehat{\rho}$ with respect to H_B is the mapping $\widehat{\rho}^A : \mathcal{L}_1(H) \longrightarrow \mathcal{L}_1(H_A)$ defined by

$$(\widehat{\rho}^A \psi_A|\psi'_A)_{H_A} = \sum_j (\widehat{\rho}(\psi_A \otimes \phi_{B,j})|\psi'_A \otimes \phi_{B,j})_H \tag{16.1}$$

where $(\phi_{B,j})_j$ is an arbitrary orthonormal basis of H_B.

Note that this definition can be written more compactly as, equivalently,

$$\widehat{\rho}^A = \sum_j \epsilon^*_{B,j} \widehat{\rho} \epsilon_{B,j} \tag{16.2}$$

where $\epsilon_{B,j}: H_A \longrightarrow H_A \otimes H_B$ is the extension map defined by $\epsilon_{B,j}\psi_A = \psi_A \otimes \phi_{B,j}$; its adjoint $\epsilon^*_{B,j}: H_A \otimes H_B \longrightarrow H_A$ is given by

$$\epsilon^*_{B,j}(\psi_A \otimes \psi_B) = (\psi_B|\phi_{B,j})_{H_B} \psi_A.$$

One easily shows that the operator $\widehat{\rho}^A_j = \epsilon^*_{B,j} \widehat{\rho} \epsilon_{B,j}$ defined by

$$(\widehat{\rho}^A_j \psi_A|\psi'_A)_{H_A} = (\widehat{\rho}(\psi_A \otimes \phi_{B,j})|\psi'_A \otimes \phi_{B,j})_H$$

is a density operator that does not depend on the choice of the orthonormal basis $(\phi_{B,j})_j$ of H_B, and that:

Proposition 207. *The partial trace $\widehat{\rho}^A = \mathrm{Tr}_B(\widehat{\rho})$ of $\widehat{\rho}$ is the unique operator $\widehat{\rho}^A = \mathrm{Tr}_A(\widehat{\rho}) \in \mathcal{L}_1(H_A)$ such that*

$$\mathrm{Tr}(\widehat{\rho}^A \widehat{Y}) = \mathrm{Tr}(\widehat{\rho}(\widehat{Y} \otimes \widehat{I}_B)) \tag{16.3}$$

for all $\widehat{Y} \in \mathcal{B}(H_A)$; here $\mathrm{Tr}(\widehat{\rho}^A \widehat{Y})$ is the trace of $\widehat{\rho}^A \widehat{Y} \in \mathcal{L}_1(H_A)$ and $\widehat{I}_B$ the identity operator on H_B.

Notice that the left hand side of (13.21) is well-defined since $\mathcal{L}_1(H_A)$ is an ideal in $\mathcal{B}(H_A)$. Similarly, $\mathrm{Tr}(\widehat{\rho}(\widehat{Y} \otimes \widehat{I}_B))$ is also well-defined since $\widehat{Y} \otimes \widehat{I}_B \in \mathcal{B}(H)$. The uniqueness of a trace class operator $\widehat{\rho}^A$ satisfying (13.21) is clear: If there were two such operators, then we would gave $\mathrm{Tr}(\widehat{\rho}(\widehat{Y} \otimes \widehat{I}_B)) = 0$ for all $Y \in \mathcal{B}(H_A)$, hence in particular, taking $\widehat{Y} = \widehat{I}_A$, $\mathrm{Tr}(\widehat{\rho}) = 0$ which is not possible. The verification of the equality (13.21) is left to the reader.

16.1.3 The reduced density operator in harmonic analysis

We now assume $H_A = L^2(\mathbb{R}^{n_A})$ and $H_B = L^2(\mathbb{R}^{n_B})$, so that we may identify the Hilbert space $H = H_A \otimes H_B$ with $L^2(\mathbb{R}^n)$. Let n_A, n_B be two positive integers such that $n = n_A + n_B$.

We identify the direct sum $\mathbb{R}^{2n_A} \oplus \mathbb{R}^{2n_B}$ with $\mathbb{R}^{2n}$ and the symplectic form σ on $\mathbb{R}^{2n}$ with $\sigma_A \oplus \sigma_B$ where σ_A (*resp.* σ_B) is the standard symplectic form on $\mathbb{R}^{2n_A}$ (*resp.* $\mathbb{R}^{2n_B}$).

Let $\widehat{\rho}$ be a density operator on $L^2(\mathbb{R}^n)$ with Wigner distribution ρ. Assume that the trace formula

$$\operatorname{Tr}(\widehat{\rho}) = \int_{\mathbb{R}^{2n}} \rho(z)dz = 1$$

holds (which is the case if, for example, $\rho \in \Gamma_\rho^m(\mathbb{R}^{2n})$ with $m < -2n$, see Proposition 157, or if $\widehat{\rho}$ is a Feichtinger state), we give the following definition:

Definition 208. The *reduced density operator* $\widehat{\rho}^A$ is

$$\widehat{\rho}^A = (2\pi\hbar)^{n_A} \operatorname{Op_W}(\rho^A) \tag{16.4}$$

where ρ^A is the function on $\mathbb{R}^{2n_A}$ defined by

$$\rho^A(z_A) = \int_{\mathbb{R}^{n_B}} \rho(z_A, z_B)dz_B. \tag{16.5}$$

Here is an existence result:

Proposition 209. *Assume that $\rho \in \Gamma_\rho^m(\mathbb{R}^{2n})$ with $m < -2n$. Then $\rho^A \in \Gamma_\rho^{m_A}(\mathbb{R}^{2n_A})$ for every $m_A < -2n_A$. The operator $\widehat{\rho}^A$ is a density operator on $L^2(\mathbb{R}^{n_A})$ with Wigner distribution ρ^A.*

Proof. The integral (16.5) is convergent in view of the trivial inequality $(1 + |z|^2)^m \leq (1 + |z_B|^2)^m$. Choosing $m_A < -2n_A$ and $m_B < -2n_B$ such that $m = m_A + m_B$, we have

$$\langle z\rangle^{m-|\alpha|} \leq \langle z_A\rangle^{m_A-|\alpha|}\langle z_B\rangle^{m_B}$$

as follows from the inequality

$$(1 + |z|^2)^{m-|\alpha|} \leq (1 + |z_A|^2)^{m_A-|\alpha|}(1 + |z_B|^2)^{m_B}.$$

Using the Shubin estimates (5.15), we thus have

$$\begin{aligned}\partial_{z_A}^\alpha \rho^A(z_A) &= \int_{\mathbb{R}^{2n_B}} \partial_{z_A}^\alpha \rho(z_A, z_B)dz_B \\ &\leq C_\alpha \langle z\rangle^{m-|\alpha|} \int_{\mathbb{R}^{2n_B}} \langle z_B\rangle^{m_B} dz_B,\end{aligned}$$

and hence $\rho^A \in \Gamma^{m_A}(\mathbb{R}^{2n_A})$ since the integral over $\mathbb{R}^{n_B}$ is convergent in view of the inequality $m_B < -2n_B$. It follows from Proposition 157 that $\widehat{\rho}^A$ is a trace class operator whose trace is

$$\operatorname{Tr}_A(\widehat{\rho}^A) = \int_{\mathbb{R}^{2n_A}} \rho^A(z_A)dz_A = 1. \tag{16.6}$$

There remains to be shown that $\widehat{\rho}^A \geq 0$. In view of Proposition 165, it is sufficient to prove that the Fourier transform $(\rho^A)_\diamond$ is continuous and satisfies $\Lambda^A_{(N)} \geq 0$ for every integer $N > 0$ where $\Lambda^A_{(N)} = (\Lambda^A_{jk})_{j,k}$ with

$$\Lambda^A_{jk} = e^{-\frac{i\hbar}{2}\sigma_A(z_{A,j},z_{A,k})}(\rho^A)_\diamond(z_{A,j} - z_{A,k}),$$

the vectors $z_{A,j}$ and $z_{A,k}$ of $\mathbb{R}^{2n_A}$ being arbitrary. The continuity of $(\rho^A)_\diamond$ being obvious (Riemann–Lebesgue Lemma), all we have to do is to show that $\Lambda^A_{(N)} \geq 0$. We first observe that, by Fubini's theorem, $(\rho^A)_\diamond(z_A) = \rho_\diamond(z_A \oplus 0)$, and hence

$$\Lambda^A_{jk} = e^{-\frac{i\hbar}{2}\sigma_A(z_{A,j}\oplus 0,z_{A,k}\oplus 0)}\rho_\diamond((z_{A,j} \oplus 0) - (z_{A,k} \oplus 0));$$

the matrix $\Lambda^A_{(N)}$ is thus the matrix $\Lambda_{(N)}$ corresponding to the particular choices $z_j = z_{A,j} \oplus 0$ and $z_k = z_{A,k} \oplus 0$. Since $\rho_\diamond$ satisfies the conditions in Proposition 165, we must have $\Lambda^A_{(N)} \geq 0$, hence $(\rho^A)_\diamond$ also satisfies these conditions. □

From now on, we will write the covariance matrix Σ in the AB-ordering as

$$\Sigma = \begin{pmatrix} \Sigma_{AA} & \Sigma_{AB} \\ \Sigma_{BA} & \Sigma_{BB} \end{pmatrix} \quad \text{with } \Sigma_{BA} = \Sigma^T_{AB}, \tag{16.7}$$

the blocks $\Sigma_{AA}, \Sigma_{AB}, \Sigma_{BA}, \Sigma_{BB}$ having dimensions $2n_A\times 2n_A$, $2n_A\times 2n_B$, $2n_B\times 2n_A$, $2n_B\times 2n_B$, respectively. In this notation, the quantum condition on the covariance matrix reads

$$\Sigma + \frac{i\hbar}{2}J_{AB} \geq 0 \tag{16.8}$$

where we have set

$$J_{AB} = J_A \oplus J_B = \begin{pmatrix} J_A & 0_{n_A\times n_B} \\ 0_{n_B\times n_A} & J_B \end{pmatrix},$$

J_A and J_B being the standard symplectic matrices of $\mathbb{R}^{2n_A}$ and $\mathbb{R}^{2n_B}$, respectively.

The covariance matrices Σ_A and Σ_B of the reduced density operators are, respectively, the blocks Σ_{AA} and Σ_{BB} of Σ as immediately follows from the definitions, using the formulas

$$\rho^A(z_A) = \int_{\mathbb{R}^{n_B}} \rho(z_A, z_B)dz_B, \quad \rho_B(z_B) = \int_{\mathbb{R}^{n_A}} \rho(z_A, z_B)dz_A.$$

These reduced covariance matrices satisfy the quantum conditions

$$\Sigma_A + \frac{i\hbar}{2}J_A \geq 0 \quad \text{and} \quad \Sigma_B + \frac{i\hbar}{2}J_B \geq 0, \tag{16.9}$$

and the covariance ellipsoids of $\widehat{\rho}^A$ and $\widehat{\rho}^B$ are, accordingly,

$$\Omega_A = \left\{z_A : \frac{1}{2}\Sigma^{-1}_{AA}z_A \cdot z_A \leq 1\right\}, \quad \Omega_B = \left\{z_B : \frac{1}{2}\Sigma^{-1}_{BB}z_B \cdot z_B \leq 1\right\}. \tag{16.10}$$

That the quantum conditions (16.9) hold follows from the fact that $\widehat{\rho}^A$ and $\widehat{\rho}^B$ are *bona fide* density operators; this can however also be seen directly by noting that the quantum condition (16.8) can be written

$$\begin{pmatrix} \Sigma_{AA} + \frac{i\hbar}{2}J_A & \Sigma_{AB} \\ \Sigma_{BA} & \Sigma_{BB} + \frac{i\hbar}{2}J_B \end{pmatrix} \geq 0.$$

16.2 Separability and the PPT condition

16.2.1 Separable and entangled quantum states

We begin by introducing some notation.

We denote by I_A the identity $(x_A, p_A) \longmapsto (x_A, p_A)$ and by $\overline{I}_B$ the involution $(x_B, p_B) \longmapsto (x_B, -p_B)$ ("partial reflection"). We set $\overline{I}_{AB} = I_A \oplus \overline{I}_B$ and $J_{AB} = J_A \oplus J_B$ where J_A (*resp.* J_B) is the standard symplectic matrix in $\mathbb{R}^{2n_A}$ (*resp.* $\mathbb{R}^{2n_B}$).

Definition 210. The density operator $\widehat{\rho}$ on $L^2(\mathbb{R}^n)$ is said to be AB-separable (or, simply: separable) if there exists a sequence $(\widehat{\rho}_j^A)_j$ (resp. $(\widehat{\rho}_j^B)_j$) of density operators on $L^2(\mathbb{R}^{n_A})$ (resp. on $L^2(\mathbb{R}^{n_B})$) and real numbers $\alpha_j \geq 0$, $\sum_j \alpha_j = 1$ such that

$$\widehat{\rho} = \sum_j \alpha_j \widehat{\rho}_j^A \otimes \widehat{\rho}_j^B, \tag{16.11}$$

the series being convergent in $\mathcal{L}_1(L^2(\mathbb{R}^n))$. If $\widehat{\rho}$ is not separable, it is said to be entangled.

It follows that the Wigner distribution of a separable state (16.11) is of the type

$$\rho(z) = \sum_j \alpha_j \rho_j^A(z_A)\rho_j^B(z_B).$$

The property of separability is not preserved by conjugation by arbitrary unitary operators. However:

Proposition 211. *Let $S_A \in \mathrm{Sp}(n_A)$ and $S_B \in \mathrm{Sp}(n_B)$, and set $S = S_A \oplus S_B$; we have $S \in \mathrm{Sp}(n)$. Let $\widehat{S} = \widehat{S}_A \oplus \widehat{S}_B \in \mathrm{Mp}(n)$ cover S. If $\widehat{\rho}$ is a separable density operator, then so is $\widehat{S}\widehat{\rho}\widehat{S}^{-1}$.*

Proof. This immediately follows from the symplectic covariance of the Weyl transform since

$$\widehat{S}(\widehat{\rho}_j^A \otimes \widehat{\rho}_j^B)\widehat{S}^{-1} = (\widehat{S}_A\widehat{\rho}_j^A\widehat{S}_A^{-1}) \otimes (\widehat{S}_B\widehat{\rho}_j^B\widehat{S}_B^{-1}).$$

The Wigner distribution of $\widehat{S}\widehat{\rho}\widehat{S}^{-1}$ is

$$\rho \circ S^{-1} = \sum_j \alpha_j (\rho_j^A \circ \widehat{S}_A^{-1}) \otimes (\rho_j^B \cdot \widehat{S}_A^{-1}).$$

□

16.2.2 The PPT theorem

We are going to prove a necessary condition for AB-separability. It makes use of the trivial equality

$$W\psi(\overline{I}_B z) = W\overline{\psi}(z_B), \tag{16.12}$$

valid for all $\psi \in L^2(\mathbb{R}^n)$. In other words, replacing $W\psi(x,p)$ with $W\psi(x,-p)$ amounts to replace $W\psi$ with $W\overline{\psi}$.

Recall from Proposition 67 in Chapter 5 that the transpose $\widehat{A}^T$ of a Weyl operator $\widehat{A} = \mathrm{Op_W}(a)$is the Weyl operator $\mathrm{Op_W}(b)$ where $b(x,p) = a(x,-p)$.

Proposition 212 (PPT). *Let $\widehat{\rho}$ be a density operator on $L^2(\mathbb{R}^n)$. Suppose that the AB-separability condition (16.11) holds. Then the partially transposed operator*

$$\widehat{\rho}^{T_B} = \sum_j \alpha_j \widehat{\rho}_j^A \otimes (\widehat{\rho}_j^B)^T, \tag{16.13}$$

that is,

$$\widehat{\rho}^{T_B} = (2\pi\hbar)^n \,\mathrm{Op_W}(\rho \circ \overline{I}_{AB}) \tag{16.14}$$

is also a density operator.

Proof. The transpose of $\widehat{\rho}_j^B$ is explicitly given by

$$(\widehat{\rho}_j^B)^T = (2\pi\hbar)^{n_B} \,\mathrm{Op_W}(\rho_j \circ \overline{I}_B),$$

hence the equivalence of (16.13) and (16.14). Suppose that the separability condition (16.11) holds; then the Wigner distribution is $\rho = \sum_j \lambda_j \rho_j^A \otimes \rho_j^B$ with

$$\rho_j^A = \sum_\ell \alpha_{j,\ell} W_A \psi_{j,\ell}^A, \quad \rho_j^B = \sum_m \beta_{j,m} W_B \psi_{j,m}^B$$

with $(\psi_{j,\ell}^A, \psi_{j,\ell}^B) \in L^2(\mathbb{R}^{n_A}) \times L^2(\mathbb{R}^{n_B})$ and $\alpha_{j,\ell}, \beta_{j,m} \geq 0$; that is

$$\rho = \sum_{j,\ell,m} \gamma_{j,\ell,m} W_A \psi_{j,\ell}^A \otimes W_B \psi_{j,m}^B$$

where $\gamma_{j,\ell,m} = \lambda_j \alpha_{j,\ell} \beta_{j,m} \geq 0$. We have

$$\rho(\overline{I}_{AB} z) = \sum_{j \in \mathcal{I}} \lambda_j \rho_j^A(z_A) \rho_j^B(\overline{I}_B z_B);$$

and thus

$$\rho \circ \overline{I}_{AB} = \sum_{j,\ell,m} \gamma_{j,\ell,m} W(\psi_{j,\ell}^A \otimes \overline{\psi}_{j,m}^B),$$

hence $\mathrm{Op_W}(\rho \circ \overline{I}_{AB})$ is also a positive semidefinite trace class operator. That we have $\mathrm{Tr}(\widehat{\rho}^{T_B}) = \mathrm{Tr}(\widehat{\rho}) = 1$ is obvious. □

The denomination PPT for the stated condition is very much used in physics; it is an acronym for *partial positive transpose.*

Proposition 20 has the following consequence. We set

$$\overline{J}_{AB} = J_A \oplus (-J_B) = \overline{I}_{AB} J_{AB} \overline{I}_{AB}$$

(hence $\overline{J}_{AB}$ is the standard symplectic matrix of $\mathbb{R}^{2n_A} \oplus \mathbb{R}^{2n_B}, \sigma_A \oplus (-\sigma_B)$).

Writing the covariance matrix in block matrix form

$$\Sigma = \begin{pmatrix} \Sigma_{AA} & \Sigma_{AB} \\ \Sigma_{BA} & \Sigma_{BB} \end{pmatrix}, \tag{16.15}$$

we have

Corollary 213. *Let $\widehat{\rho} = (2\pi\hbar)^n \operatorname{Op_W}(\rho)$ be a separable density operator. Then, in addition to the total quantum condition (16.8), we have*

$$\Sigma + \frac{i\hbar}{2}\overline{J}_{AB} \geq 0; \tag{16.16}$$

equivalently,

$$\overline{\Sigma} + \frac{i\hbar}{2} J_{AB} \geq 0 \tag{16.17}$$

where $\overline{\Sigma} = \overline{I}_{AB}\Sigma\overline{I}_{AB}$, that is

$$\overline{\Sigma} = \begin{pmatrix} \Sigma_{AA} & \Sigma_{AB}\overline{I}_B \\ \overline{I}_B\Sigma_{BA} & \overline{I}_B\Sigma_{BB}\overline{I}_B \end{pmatrix}. \tag{16.18}$$

Proof. Replacing ρ with $\rho \circ \overline{I}_{AB}$ the matrix Σ_{AA} in (16.7) remains unchanged, while Σ_{BB}, Σ_{AB} and Σ_{BA} become $\overline{I}_B\Sigma_{BB}\overline{I}_B$, $\Sigma_{AB}\overline{I}_B$ and $\overline{I}_B\Sigma_{BA}$, respectively. The covariance matrix (16.7) thus becomes $\overline{\Sigma} = \overline{I}_{AB}\Sigma\overline{I}_{AB}$. In view of Proposition 20, the operator $(2\pi\hbar)^n \operatorname{Op_W}(\rho \circ \overline{I}_B)$ is also positive semidefinite, hence we must have $\overline{\Sigma} + \frac{i\hbar}{2} J_{AB} \geq 0$ which is equivalent to $\Sigma + \frac{i\hbar}{2}\overline{I}_{AB}J_{AB}\overline{I}_{AB} \geq 0$. Since $\overline{I}_{AB}J_{AB}\overline{I}_{AB} = \overline{J}_{AB}$, this is the same thing as (16.16). □

16.2.3 The Schur complement

Let M be a $2n \times 2n$ matrix, partitioned as

$$M = \begin{pmatrix} M_{AA} & M_{AB} \\ M_{BA} & M_{BB} \end{pmatrix} \tag{16.19}$$

where the blocks $M_{AA}, M_{AB}, M_{BA}, M_{BB}$ have dimensions $2n_A \times 2n_A$, $2n_A \times 2n_B$, $2n_B \times 2n_A$, $2n_B \times 2n_B$, respectively.

Assume that is M is symmetric and positive definite, which we write $M > 0$. The symmetric matrices

$$M/M_{BB} = M_{AA} - M_{AB}M_{BB}^{-1}M_{BA} \tag{16.20}$$

$$M/M_{AA} = M_{BB} - M_{BA}M_{AA}^{-1}M_{AB} \tag{16.21}$$

are called the Schur complements of the blocks M_{BB} and M_{AA}, respectively. The positivity of M implies that M_{AA} and M_{BB} are themselves > 0 and hence invertible.

Using the obvious factorization

$$M = \begin{pmatrix} I_A & M_{AB}M_{BB}^{-1} \\ 0 & I_B \end{pmatrix} \begin{pmatrix} M/M_{BB} & 0 \\ 0 & M_{BB} \end{pmatrix} \begin{pmatrix} I_A & 0 \\ M_{BB}^{-1}M_{BA} & I_B \end{pmatrix}, \tag{16.22}$$

we readily obtain various formulas for the inverse of M; the one we will use here is

$$M^{-1} = \begin{pmatrix} (M/M_{BB})^{-1} & -(M/M_{BB})^{-1}M_{AB}M_{BB}^{-1} \\ -M_{BB}^{-1}M_{BA}(M/M_{BB})^{-1} & (M/M_{AA})^{-1} \end{pmatrix}. \tag{16.23}$$

Also note that it immediately follows from (16.22) that

$$\det M = \det(M/M_{BB}) \det M_{BB} \tag{16.24}$$

and

$$\det M = \det(M/M_{AA}) \det M_{AA}. \tag{16.25}$$

16.2.4 Orthogonal projections of the covariance ellipsoid

We are going to show (Proposition 215) that the orthogonal projections of the full covariance ellipsoid Ω on $\mathbb{R}^{2n_A}$ and $\mathbb{R}^{2n_B}$ are the covariance ellipsoids of the reduced density operators. The following is a restatement of Lemma 202 in our present notation:

Lemma 214. *Let Π_A (resp. Π_B) be the orthogonal projection $\mathbb{R}^{2n} \longrightarrow \mathbb{R}^{2n_A}$ (resp. $\mathbb{R}^{2n} \longrightarrow \mathbb{R}^{2n_B}$) and Ω_R the phase space ellipsoid*

$$\Omega_R = \{z \in \mathbb{R}^{2n} : Mz^2 \leq R^2\}$$

($R > 0$). We have

$$\Pi_A\Omega_R = \{z_A \in \mathbb{R}^{2n_A} : (M/M_{BB})z_A^2 \leq R^2\} \tag{16.26}$$

$$\Pi_B\Omega_R = \{z_B \in \mathbb{R}^{2n_B} : (M/M_{AA})z_B^2 \leq R^2\}. \tag{16.27}$$

It follows from Lemma 214 that the orthogonal projections on $\mathbb{R}^{2n_A}$ and $\mathbb{R}^{2n_B}$ of the covariance ellipsoid Ω of $\widehat{\rho}$ are just the covariance ellipsoids of the reduced operators $\widehat{\rho}^A$ and $\widehat{\rho}^B$:

Proposition 215. *The covariance ellipsoids Ω_A and Ω_B of the reduced quantum states $\widehat{\rho}^A$ and $\widehat{\rho}^B$ are the orthogonal projections on $\mathbb{R}^{2n_A}$ and $\mathbb{R}^{2n_B}$ of the covariance ellipsoid Ω of $\widehat{\rho}$:*

$$\Omega_A = \Pi_A\Omega = \left\{z_A \in \mathbb{R}^{2n_A} : \frac{1}{2}\Sigma_{AA}^{-1}z_A \cdot z_A \leq 1\right\} \tag{16.28}$$

$$\Omega_B = \Pi_B\Omega = \left\{z_B \in \mathbb{R}^{2n_B} : \frac{1}{2}\Sigma_{BB}^{-1}z_B \cdot z_B \leq 1\right\}. \tag{16.29}$$

Proof. Let $M = \frac{\hbar}{2}\Sigma^{-1}$; the covariance matrix is then defined by the inequality $Mz \cdot z \leq \hbar$ so that its projections $\Pi_A\Omega$ and $\Pi_B\Omega$ are defined by

$$(M/M_{BB})z_A \cdot z_A \leq \hbar, \quad (M/M_{AA})z_B \cdot z_B \leq \hbar$$

in view of Lemma 214 with $R = \sqrt{\hbar}$. Writing M in block-matrix form (16.19), its inverse has the form

$$M^{-1} = \begin{pmatrix} (M/M_{BB})^{-1} & * \\ * & (M/M_{AA})^{-1} \end{pmatrix} \tag{16.30}$$

(formula (16.23)), and hence

$$(M/M_{BB})^{-1} = \frac{\hbar}{2}\Sigma_{AA} \quad \text{and} \quad (M/M_{AA})^{-1} = \frac{\hbar}{2}\Sigma_{BB}.$$

Formulas (16.28) and (16.29). □

16.3 Werner and Wolf's condition

16.3.1 Statement and equivalent formulation

The following crucial necessary condition for AB-separability was proven by Werner and Wolf:

Proposition 216 (Werner and Wolf). *Suppose that the density operator $\widehat{\rho}$ with covariance matrix Σ is AB-separable. Then there exist two real symmetric matrices Σ_A and Σ_B of dimensions $2n_A \times 2n_A$ and $2n_B \times 2n_B$, respectively, satisfying the quantum conditions*

$$\Sigma_A + \frac{i\hbar}{2}J_A \geq 0 \quad \textit{and} \quad \Sigma_B + \frac{i\hbar}{2}J_B \geq 0 \tag{16.31}$$

and such that

$$\Sigma \geq \Sigma_A \oplus \Sigma_B. \tag{16.32}$$

We omit the proof which is quite technical, and instead show that Werner and Wolf's result can be considerably refined using the properties of the symplectic group. We recall that the quantum condition $\Sigma + \frac{i\hbar}{2}J \geq 0$ on a covariance matrix is equivalent to the following property: Ω *contains a quantum blob* (Theorem 201).

Proposition 217. *The Werner–Wolf condition (16.32) is equivalent to the existence of two positive definite symplectic matrices*

$$P_A = (S_A^T S_A)^{-1} \in \mathrm{Sp}(n_A), \quad P_B = (S_B^T S_B)^{-1} \in \mathrm{Sp}(n_B) \tag{16.33}$$

($S_A \in \mathrm{Sp}(n_A)$ and $S_B \in \mathrm{Sp}(n_B)$) such that

$$\Sigma \geq \frac{\hbar}{2}(P_A \oplus P_B). \tag{16.34}$$

Equivalently, the covariance ellipsoid Ω contains a quantum blob $(S_A \oplus S_B)(B^{2n}(\sqrt{\hbar}))$.

Proof. The sufficiency of the condition is clear since $\Sigma_A = \frac{\hbar}{2}P_A$ and $\Sigma_A = \frac{\hbar}{2}P_A$ satisfy the conditions (16.31). Assume conversely that $\Sigma \geq \Sigma_A \oplus \Sigma_B$. In view of Williamson's diagonalization theorem, there exist $S_A \in \mathrm{Sp}(n_A)$ and $S_B \in \mathrm{Sp}(n_B)$ such that $S_A \Sigma_A S_A^T = D_A$ and $S_B \Sigma_B S_B^T = D_B$ where

$$D_A = \begin{pmatrix} \Lambda_A & 0 \\ 0 & \Lambda_A \end{pmatrix}, \quad D_B = \begin{pmatrix} \Lambda_B & 0 \\ 0 & \Lambda_B \end{pmatrix}$$

and Λ_A, Λ_B being the diagonal matrices consisting of the symplectic eigenvalues $\lambda_1^A, \dots, \lambda_{n_A}^A$ of Σ_A and $\lambda_1^B, \dots, \lambda_{n_B}^B$ of Σ_B. Since $S_A J_A S_A^T = J_A$ and $S_B J_B S_B^T = J_B$, the conditions $\Sigma_A + \frac{i\hbar}{2}J_A \geq 0$ and $\Sigma_B + \frac{i\hbar}{2}J_B \geq 0$ are equivalent to $D_A + \frac{i\hbar}{2}J_A \geq 0$ and $D_B + \frac{i\hbar}{2}J_B \geq 0$. These conditions imply in turn that $D_A \geq \frac{\hbar}{2}I_A$ and $D_B \geq \frac{\hbar}{2}I_B$: the characteristic equation of $D_A + \frac{i\hbar}{2}J_A$ is

$$\det\left((\Lambda_A - \lambda I_A)^2 - \frac{1}{4}\hbar^2 I_A \right) = 0,$$

and, writing $\Lambda_A = \mathrm{diag}(\lambda_1^A, \dots, \lambda_n^A)$, this equation is equivalent to the set of equations

$$(\lambda_j^A - \lambda)^2 - \frac{1}{4}\hbar^2 = 0, \quad 1 \leq j \leq n,$$

whose solutions are the real numbers $\lambda_j = \lambda_j^A \pm \frac{\hbar}{2}$. Since $\lambda_j \geq 0$, we must thus have $\lambda_j^A \geq \frac{\hbar}{2}$, and hence $D_A \geq \frac{\hbar}{2}I_A$. Similarly, $D_B \geq \frac{\hbar}{2}I_B$, so we must have the inequalities

$$\Sigma_A = S_A^{-1} D_A (S_A^T)^{-1} \geq (S_A^T S_A)^{-1}$$
$$\Sigma_B = S_B^{-1} D_A (S_B^T)^{-1} \geq (S_B^T S_B)^{-1}.$$

Setting $P_A = (S_A^T S_A)^{-1}$ and $P_B = (S_B^T S_B)^{-1}$, the inequality (16.34) follows. □

16.3.2 A geometric consequence

The Werner and Wolf conditions (16.31) mean that Σ_A and Σ_B are quantum covariances matrices, hence the sum

$$\Sigma_A \oplus \Sigma_B \equiv \begin{pmatrix} \Sigma_A & 0 \\ 0 & \Sigma_B \end{pmatrix}$$

is a quantum covariance matrix in its own right. It follows from the condition (16.32) that the corresponding covariance ellipsoid, which we denote as

$$\Omega_{A\oplus B} = \left\{ z_A \oplus z_B : \frac{1}{2}\Sigma_A^{-1} z_A^2 + \frac{1}{2}\Sigma_B^{-1} z_B^2 \leq 1 \right\}, \tag{16.35}$$

is included in Ω. Moreover, the ellipsoid $\Omega_{A\oplus B}$ always contains a quantum blob of the form

$$\begin{aligned} \Omega_{AB} &= (S_A \oplus S_B)(B^{2n}(\sqrt{\hbar})) && (16.36) \\ &= \{z_A \oplus z_B : |S_A^{-1} z_A|^2 + |S_B^{-1} z_B|^2 \leq \hbar\}. && (16.37) \end{aligned}$$

Hence, if the density operator $\widehat{\rho}$ with covariance ellipsoid Ω is separable, then there exist quantum covariance ellipsoids of the form (16.35) and (16.36) such that the following inclusions hold

$$\Omega \supset \Omega_{A\oplus B} \supset \Omega_{AB}. \tag{16.38}$$

This result has interesting consequences for the covariance ellipsoids

$$\Omega_A = \left\{ z_A : \frac{1}{2}\Sigma_{AA}^{-1} z_A^2 \leq 1 \right\} \quad \text{and} \quad \Omega_B = \left\{ z_B : \frac{1}{2}\Sigma_{BB}^{-1} z_B^2 \leq 1 \right\}$$

of the reduced density operators $\widehat{\rho}_A$ and $\widehat{\rho}_B$. We first show that:

Proposition 218. *The orthogonal projections $\Pi_A\Omega_{AB}$ and $\Pi_B\Omega_{AB}$ of Ω_{AB} onto $\mathbb{R}^{2n_A}$ and $\mathbb{R}^{2n_B}$ satisfy*

$$\Pi_A\Omega_{AB} = S_A(B^{2n_A}(\sqrt{\hbar})), \quad \Pi_B\Omega_{AB} = S_B(B^{2n_B}(\sqrt{\hbar})). \tag{16.39}$$

Proof. This result is easily proved directly from the definition of Ω_{AB}. □

Finally, from (16.38), we easily conclude that the covariant ellipsoids Ω_A and Ω_B impose the following constraints on the symplectic matrices S_A and S_B of Proposition 21:

Corollary 219. *Assume that the density operator $\widehat{\rho}$ with covariant ellipsoid Ω is separable. Then the symplectic matrices S_A and S_B of Proposition 21 satisfy:*

$$S_A B^{2n_A}(\sqrt{\hbar}) \subset \Omega_A, \quad S_B B^{2n_B}(\sqrt{\hbar}) \subset \Omega_B \tag{16.40}$$

Proof. From (16.38), we have $\Omega_{AB} \subset \Omega$ and so:

$$\Pi_A \Omega_{AB} \subset \Pi_A \Omega = \Omega_A, \quad \Pi_B \Omega_{AB} \subset \Pi_B \Omega = \Omega_B. \qquad \square$$

16.4 Comments and references

The study of the notions of separability and entanglement can be historically traced back to the famous "EPR paper" [24] by Einstein, Rosen, and Podolsky, where the apparent paradoxical predictions of quantum mechanics about strongly correlated systems were first discussed. It was Schrödinger who coined the word entanglement (*Verschränkung* in German) in a letter to Einstein.

There is an immense physical literature on entanglement and separability, still growing every day. For recent developments, see Lami et al. [52] and Serafini [65].

The PPT condition was first precisely stated (in physicist's language) by the Horodecki family in [43, 44] and by Asher Peres [62]. The Werner and Wolf theorem was proven in [74] following earlier constructions by Reinhardt Werner [73]. For useful formulas of the Schur complement, see Zhang [77].

Part of the last section contains material that is developed in a paper which is joint work [20] with Nuno Dias and Joao Prata.

A numerical approach to the problem of entanglement can be found in the paper by Leinaas et al. [53].

We stress that the important PPT and Werner–Wolf theorems are necessary but not sufficient conditions; at the time of writing, a tractable sufficient condition is still lacking outside of a few particular cases (for instance, the Werner–Wolf condition is not only necessary but also sufficient for Gaussian state, as we will see in next chapter).

17 Separability of Gaussian states

Gaussian quantum states, i. e., states whose Wigner distribution is a Gaussian function, play a very privileged role in quantum mechanics. Not only are they somewhat easier to study analytically than more general states, but they are also omnipresent in all applications (quantum optics and computing, for instance). They have, in addition, very interesting properties; for instance, the quantum condition on the covariance matrix is sufficient to ensure "quantumness", and so is the Werner–Wolf condition for separability.

17.1 The purity of a Gaussian state

Consider now, as we have done several times before in this book, a non-degenerate real Gaussian function on $\mathbb{R}^{2n}$ of the type

$$\rho(z) = \frac{1}{(2\pi)^n\sqrt{\det\Sigma}}e^{-\frac{1}{2}\Sigma^{-1}(z-\bar{z})^2}$$

centered at $\bar{z} \in \mathbb{R}^{2n}$, where Σ is a positive definite real symmetric $2n \times 2n$ matrix (the "covariance matrix"). We will only consider the case $\bar{z} = 0$; the more general case is easily reduced to the former by a phase-space translation. Hence we assume that

$$\rho(z) = \frac{1}{(2\pi)^n\sqrt{\det\Sigma}}e^{-\frac{1}{2}\Sigma^{-1}z^2} \tag{17.1}$$

where Σ is a real positive definite symmetric $2n \times 2n$ matrix. This function represents in all cases a probability distribution since $\rho \geq 0$ and is normalized

$$\int_{\mathbb{R}^{2n}} \rho(z)dz = 1$$

as easily checked by diagonalizing Σ and using the elementary formula

$$\frac{1}{\sigma\sqrt{2\pi}}\int_{-\infty}^{\infty} e^{-x^2/2\sigma^2}dx = 1.$$

17.1.1 A general purity formula for Gaussian sates

We know from Proposition 168 in Chapter 13 that the function ρ defined by (17.1) is the Wigner distribution of a density operator if and only if the quantum condition

$$\Sigma + \frac{i\hbar}{2}J \geq 0$$

is satisfied. Assuming that this condition is satisfied, we have:

https://doi.org/10.1515/9783110722772-017

Proposition 220. *Let $\widehat{\rho}$ be the density matrix with Gaussian Wigner distribution defined previously. The purity of $\widehat{\rho}$ is*

$$\mu(\widehat{\rho}) = \left(\frac{\hbar}{2}\right)^n \det(\Sigma^{-1/2}). \tag{17.2}$$

Proof. We have (formula (14.8))

$$\operatorname{Tr}(\widehat{\rho}^2) = (2\pi\hbar)^n \int_{\mathbb{R}^{2n}} \rho^2(z)dz.$$

Now, by definition of the Gaussian ρ,

$$\int_{\mathbb{R}^{2n}} \rho^2(z)dz = \left(\frac{1}{2\pi}\right)^{2n} (\det\Sigma)^{-1} \int_{\mathbb{R}^{2n}} e^{-\Sigma^{-1}z^2} dz,$$

hence, using the elementary formula from the theory of Gaussian integrals,

$$\int_{\mathbb{R}^{2n}} e^{-Mz^2} dz = \pi^n (\det M)^{-1/2},$$

valid for every positive-definite symmetric matrix M, we get

$$\int_{\mathbb{R}^{2n}} \rho^2(z)dz = \left(\frac{1}{4\pi}\right)^n (\det\Sigma)^{1/2};$$

formula (17.2) follows. □

Notice that, since $\operatorname{Tr}(\widehat{\rho}^2) \leq 1$, we must have $\det\Sigma \geq (\frac{1}{2}\hbar)^n$ and $\mu(\widehat{\rho}) = 1$ if and only if $\det(\Sigma) = (\frac{1}{2}\hbar)^n$.

17.1.2 Pure Gaussian states

Let us return to the generalized Gaussians studied in Chapter 4. We slightly change the notation and write $\psi_{X,Y}$ instead of ψ_M to avoid confusion. Let X and Y be real symmetric $n \times n$ matrices, with $X > 0$. To these matrices, we associate the Gaussian function $\phi_{X,Y}$ on $\mathbb{R}^n$ defined by

$$\psi_{X,Y}(x) = (\pi\hbar)^{-n/4} (\det X)^{1/4} e^{-\frac{1}{2\hbar}(X+iY)x^2} \tag{17.3}$$

where we are writing $(X + iY)x^2$ for $(X + iY)x \cdot x$. The function $\psi_{X,Y}$ is L^2-normalized $\|\psi_{X,Y}\|_{L^2} = 1$, and its Wigner transform is given by

$$W\upsilon_{X,Y}(z) = (\pi\hbar)^{-n} e^{-\frac{1}{\hbar}S^T Sz^2} \tag{17.4}$$

where

$$S = \begin{pmatrix} X^{1/2} & 0 \\ X^{-1/2}Y & X^{-1/2} \end{pmatrix} \tag{17.5}$$

is symplectic, so that

$$S^T S = \begin{pmatrix} X + YX^{-1}Y & YX^{-1} \\ X^{-1}Y & X^{-1} \end{pmatrix}. \tag{17.6}$$

Setting $\Sigma^{-1} = \frac{\hbar}{2} S^T S$, we can rewrite the Wigner transform (17.4) in the form

$$W\psi_{X,Y}(z) = \frac{1}{(2\pi)^n \sqrt{\det \Sigma}} e^{-\frac{1}{2}\Sigma^{-1}z^2},$$

hence to $\rho_{X,Y} = W\psi_{X,Y}$ corresponds a Gaussian density operator $\widehat{\rho}_{X,Y}$. The quantum condition on the covariance matrix becomes here $S^T S + iJ \geq 0$; since $(S^T)^{-1} J S^{-1} = J$, this is equivalent to $I + iJ \geq 0$ (which is trivially satisfied).

Proposition 221. *The only pure Gaussian states are those with Wigner distribution* $W\psi_{X,Y}$.

Proof. Let $\Sigma = S'^T D S'$ be a Williamson diagonalization of Σ. We have

$$\det(\Sigma) = \det(J\Sigma) = \lambda_1^2 \cdots \lambda_n^2$$

where the λ_j are the symplectic eigenvalues of Σ. Since we must have $\lambda_j \leq \frac{1}{2}\hbar$ (Proposition 168), the equality $\det(\Sigma) = (\frac{1}{2}\hbar)^n$ requires that we must have $\lambda_j = \frac{1}{2}\hbar$ for all $j = 1, 2, \ldots, n$. In this case, the matrix D in the Williamson diagonalization $\Sigma = S'^T D S'$ is the identity so that we must have $\Sigma = S^T S$, hence

$$\rho(z) = \frac{1}{(2\pi)^n \sqrt{\det \Sigma}} e^{-\frac{1}{2}(S'^T S')^{-1} z^2},$$

and the result follows from the previous discussion, setting $S = (S'^T)^{-1}$. □

17.2 Separability of Gaussian quantum states

17.2.1 Generalities, a sufficient condition for separability

Notice that Proposition 218 gives us an alternative proof of the fact that the partial trace operators $\widehat{\rho}^A$ and $\widehat{\rho}^B$ are density operators in the Gaussian case. In fact, to prove this, we had to use for the general case the KLM conditions (Proposition 165) to prove the positivity properties $\widehat{\rho}^A \geq 0$ and $\widehat{\rho}^B \geq 0$. In the Gaussian case, the equalities (16.39) imply that

$$\Sigma_A + \frac{i\hbar}{2} J_A \geq 0, \quad \Sigma_B + \frac{i\hbar}{2} J_B \geq 0, \tag{17.7}$$

hence $\widehat{\rho}^A$ and $\widehat{\rho}^B$ are (Gaussian) density operators. Recall that the purity of $\widehat{\rho}$ is given by

$$\mu(\widehat{\rho}) = \left(\frac{\hbar}{2}\right)^n (\det \Sigma)^{-1/2} = \sqrt{\det M}. \tag{17.8}$$

That the terminology "covariance matrix" applied to Σ is justified in the quantum case because it is, in classical statistical mechanics, clear, as is the fact that we have $\rho \in \Gamma^m(\mathbb{R}^{2n})$ for every $m < -2n$, hence $\rho^A \in \Gamma^{m_A}(\mathbb{R}^{2n_A})$ for every $m_A < -2n_A$.

17.2.2 Reduced states of a Gaussian density operator

Let us describe in detail the reduced states of a Gaussian quantum state. We introduce the notation

$$M = \frac{\hbar}{2}\Sigma^{-1}.$$

Notice that we have the equivalence

$$\Sigma + \frac{i\hbar}{2}J \geq 0 \Longleftrightarrow M^{-1} + iJ \geq 0.$$

We will write M in block matrix form

$$M = \begin{pmatrix} M_{AA} & M_{AB} \\ M_{BA} & M_{BB} \end{pmatrix}, \quad M_{AB} = M_{BA}^T.$$

Recall that the Schur complements M/M_{BB} and M/M_{AA} are symmetric positive definite and given by

$$M/M_{BB} = M_{AA} - M_{AB}M_{BB}^{-1}M_{BA} \tag{17.9}$$

$$M/M_{AA} = M_{BB} - M_{BA}M_{AA}^{-1}M_{AB}. \tag{17.10}$$

Proposition 222. *The reduced density operator $\widehat{\rho}^A$ is a Gaussian state with Wigner distribution*

$$\rho^A(z_A) = (\pi\hbar)^{-n_A}(\det M/M_{BB})^{1/2}e^{-\frac{1}{\hbar}(M/M_{BB})z_A^2}; \tag{17.11}$$

and its covariance ellipsoid

$$\Omega_A = \{z_A : (M/M_{BB})z_A^2 \leq \hbar\} \tag{17.12}$$

is the orthogonal projection $\Pi_A\Omega$ on $\mathbb{R}^{2n_A}$ of the covariance ellipsoid Ω of $\widehat{\rho}$.

Proof. Writing $z = z_A \oplus z_B$, we have

$$Mz^2 = M_{AA}z_A^2 + 2M_{BA}z_A \cdot z_B + M_{BB}z_B^2,$$

so that

$$\int_{\mathbb{R}^{2n_B}} e^{-\frac{1}{\hbar}Mz^2} dz_B = e^{-\frac{1}{\hbar}M_{AA}z_A^2} \int_{\mathbb{R}^{2n_B}} e^{-\frac{1}{\hbar}(M_{BB}z_B^2+2M_{BA}z_A \cdot z_B)} dz_B.$$

Setting $z_B = u_B - M_{BB}^{-1}M_{BA}z_A$, we have

$$M_{BB}z_B^2 + 2M_{BA}z_A \cdot z_B = M_{BB}u_B^2 - M_{AB}M_{BB}^{-1}M_{BA}z_A^2,$$

and hence, integrating with respect to the variables z_B,

$$\int_{\mathbb{R}^{2n_B}} e^{-\frac{1}{\hbar}Mz^2} dz_B = e^{-\frac{1}{\hbar}(M_{AA}-M_{AB}M_{BB}^{-1}M_{BA})z_A^2} \int_{\mathbb{R}^{2n_B}} e^{-\frac{1}{\hbar}M_{BB}u_B^2} du_B.$$

Using the classical formula

$$\int_{\mathbb{R}^{2n_B}} e^{-\frac{1}{\hbar}M_{BB}u_B^2} du_B = (\pi\hbar)^{n_B}(\det M_{BB})^{-1/2},$$

we thus have

$$\int_{\mathbb{R}^{2n_B}} e^{-\frac{1}{\hbar}Mz^2} dz_B = (\pi\hbar)^{n_B}(\det M_{BB})^{-1/2} e^{-\frac{1}{\hbar}(M/M_{BB})z_A^2}$$

where M/M_{BB} is the Schur complement of M_{BB} of M; the identity (17.11) now follows from formula (16.24). The covariance ellipsoid of the reduced state $\widehat{\rho}^A$ is given by (17.12), and, in view of Lemma 214, it is indeed the orthogonal projection $\Pi_A\Omega$ of Ω on $\mathbb{R}^{2n_A}$. □

Corollary 223. *The purity of the reduced density operator* $\widehat{\rho}^A$ *is*

$$\mu(\widehat{\rho}^A) = (\det M/M_{BB})^{1/2}, \tag{17.13}$$

and $\widehat{\rho}^A$ *is a pure state if and only if* $M/M_{BB} \in \mathrm{Sp}(n_A)$, *in which case we have* $\mu(\widehat{\rho}) = \det M_{BB}$.

Proof. In view of formulas (17.8) and (17.11), the purity of $\widehat{\rho}^A$ is given by (17.13). It follows that $\mu(\widehat{\rho}^A) = 1$ if and only if $\det M/M_{BB} = 1$; by the same token as used in the proof of Lemma 221, we must then have $M/M_{BB} \in \mathrm{Sp}(n_A)$. The equality $\mu(\widehat{\rho}) = \det M_{BB}$ follows from the identity (16.24). □

17.2.3 Back to the Werner and Wolf conditions

It turns out that the Werner and Wolf's conditions in Proposition 216 are sufficient for a Gaussian state to be separable:

Proposition 224. *Assume that there exist two partial covariance matrices Σ_A and Σ_B satisfying the quantum conditions (17.7) and such that*

$$\Sigma \geq \Sigma_A \oplus \Sigma_B. \tag{17.14}$$

Then the Gaussian state

$$\rho(z) = \left(\frac{1}{2\pi}\right)^n \frac{1}{\sqrt{\det \Sigma}} e^{-\frac{1}{2}\Sigma^{-1}z^2}$$

is separable.

The conditions (17.7) mean that Σ_A and Σ_B are quantum covariances matrices, hence the sum

$$\Sigma_A \oplus \Sigma_B \equiv \begin{pmatrix} \Sigma_A & 0 \\ 0 & \Sigma_A \end{pmatrix}$$

is a quantum covariance matrix in its own right. The corresponding covariance ellipsoid, which we denote

$$\Omega_{A\oplus B} = \left\{ z_A \oplus z_B : \frac{1}{2}\Sigma_A^{-1}z_A^2 + \frac{1}{2}\Sigma_B^{-1}z_A^2 \leq 1 \right\},$$

is included in Ω, and its orthogonal projections on $\mathbb{R}^{2n_A}$ and $\mathbb{R}^{2n_B}$ are just the intersections of $\Omega_{A\oplus B}$ with the hyperplanes $z_B = 0$ and $z_A = 0$, respectively.

17.3 Disentanglement of Gaussian states

Since the action of the metaplectic group $\mathrm{Mp}(n)$ on the set of all centered Gaussians $\psi_{X,Y}$ is transitive, the previous lemma can be rephrased by saying that every pure Gaussian state is obtained from the standard Gaussian $\phi_0(x) = (\pi\hbar)^{-n/4}e^{-|x|^2/2\hbar}$ by a unitary transformation $\widehat{S} \in \mathrm{Mp}(n)$. We are in fact going to prove much more, namely that every Gaussian state (pure or mixed) can be "disentangled" using a metaplectic transform associated with a symplectic rotation.

17.3.1 A diagonalization result for positive symplectic matrices

A symplectic and positive definite matrix can be diagonalized using a symplectic rotation that is an element of $U(n) = \mathrm{Sp}(n) \cap O(2n)$.

Proposition 225. *Let $S \in \mathrm{Sp}(n)$ be positive definite and symmetric. Let $\lambda_1 \leq \cdots \leq \lambda_n \leq 1$ be the n smallest eigenvalues of S and set*

$$\Lambda = \mathrm{diag}(\lambda_1, \dots, \lambda_n; 1/\lambda_1, \dots, 1/\lambda_n). \tag{17.15}$$

There exists $U \in U(n)$ such that $S = U^T \Lambda U$.

The eigenvalues occur in pairs $(\lambda, 1/\lambda)$ of positive numbers (Proposition 11); if $\lambda_1 \leq \cdots \leq \lambda_n$ are the n first eigenvalues, then $/\lambda_1, \dots, 1/\lambda_n$ are the other n eigenvalues. Let now U be an orthogonal matrix such that $S = U^T \Lambda U$, with Λ being given by (17.15). We claim that we can choose $U \in U(n)$. It suffices to show that we can write U in the block-matrix form

$$U = \begin{pmatrix} A & -B \\ B & A \end{pmatrix}$$

with the conditions

$$AB^T = B^T A, \quad AA^T + BB^T = I. \tag{17.16}$$

Let $e_1, \dots, e_n$ be n orthonormal eigenvectors of U corresponding to the eigenvalues $\lambda_1, \dots, \lambda_n$. Since $SJ = JS^{-1}$ (because S is both symplectic and symmetric), we have, for $1 \leq k \leq n$,

$$SJe_k = JS^{-1}e_k = \frac{1}{\lambda_j} Je_k,$$

hence $\pm Je_1, \dots, \pm Je_n$ are the orthonormal eigenvectors of U corresponding to the remaining n eigenvalues $1/\lambda_1, \dots, 1/\lambda_n$. Viewing the e_k as column vectors, we write the $2n \times n$ matrix $(e_1, \dots, e_n)$ as

$$(e_1, \dots, e_n) = \begin{pmatrix} A \\ B \end{pmatrix}$$

where A and B are $n \times n$ matrices; we have

$$(-Je_1, \dots, -Je_n) = -J \begin{pmatrix} A \\ B \end{pmatrix} = \begin{pmatrix} -B \\ A \end{pmatrix},$$

hence U is indeed of the type

$$U = (e_1, \dots, e_n; -Je_1, \dots, -Je_n) = \begin{pmatrix} A & -B \\ B & A \end{pmatrix}.$$

The symplectic conditions (17.16) are automatically satisfied since $U^T U = UU^T = I$ because the diagonalizing matrix U is assumed to be orthogonal.

17.3.2 The disentanglement result

Let us state and prove our “disentanglement” result.

Proposition 226. *Let $\widehat{\rho}$ be a bipartite Gaussian density operator. There exists a symplectic rotation $U \in U(n)$ such that $\widehat{U}\widehat{\rho}\widehat{U}^{-1}$ is separable where $\widehat{U} \in \mathrm{Mp}(n)$ is any one of the two metaplectic operators covering U.*

Proof. Let $S = PR$ be the symplectic polar decomposition of $S \in \mathrm{Sp}(n)$, that is, $P \in \mathrm{Sp}(n)$, $P > 0$, and $R \in U(n)$. We have $SB^{2n}(\sqrt{\hbar}) = PB^{2n}(\sqrt{\hbar})$ since $RB^{2n}(\sqrt{\hbar}) = B^{2n}(\sqrt{\hbar})$ by rotational symmetry. In view of Proposition 225, there exists a symplectic rotation $U \in U(n)$ such that

$$P = U^T \Delta U \tag{17.17}$$

where $\Delta \in \mathrm{Sp}(n)$ is a diagonal matrix (17.15). The inclusion $SB^{2n}(\sqrt{\hbar}) \subset \Omega$ is thus equivalent to $\Delta B^{2n}(\sqrt{\hbar}) \subset U(\Omega_\Sigma)$, that is, to

$$\Delta B^{2n}(\sqrt{\hbar}) \subset \Omega_{\Sigma_U} \tag{17.18}$$

where $\Sigma_U = U\Sigma U^T$. This inclusion is equivalent to the matrix inequality

$$\frac{\hbar}{2}\Delta^2 \leq \Sigma_U. \tag{17.19}$$

We next note that Σ_U is the covariance matrix of the density operator $\widehat{\rho}_U$ with Wigner distribution $W_{\widehat{\rho}U}(z) = W_{\widehat{\rho}}(U^T z)$, that is,

$$W_{\widehat{\rho}U}(z) = \frac{1}{(2\pi)^n \sqrt{\det U\Sigma U^T}} e^{-\frac{1}{2}\Sigma^{-1}U^T z \cdot U^T z}. \tag{17.20}$$

Applying the symplectic covariance principle for Weyl operators to $\widehat{\rho}$ yields, since $U^T = U^{-1}$,

$$\widehat{\rho}_U = \widehat{U}\widehat{\rho}\widehat{U}^{-1} \tag{17.21}$$

where $\widehat{U} \in \mathrm{Mp}(n)$ is any of the two metaplectic operators $\pm\widehat{U}$ covering U. We claim that $\widehat{\rho}_U$ is separable. To see this, let us come back to the diagonal matrix Δ. It is given by

$$\Delta = \mathrm{diag}(\lambda_1, \lambda_1^{-1}, \lambda_2, \lambda_2^{-1}, \ldots, \lambda_n, \lambda_n^{-1}), \tag{17.22}$$

and, in the AB-ordering, it has the form $\Delta = \Delta_A \oplus \Delta_B$ with

$$\Delta_A = \bigoplus_{k=1}^{n_A} \Delta_k, \quad \Delta_B = \bigoplus_{k=n_A+1}^{n} \Delta_k \tag{17.23}$$

and

$$\Delta_k = \begin{pmatrix} \lambda_k & 0 \\ 0 & \lambda_k^{-1} \end{pmatrix}, \quad k = 1, \dots, n. \tag{17.24}$$

Clearly $\Delta_A \in \mathrm{Sp}(n_A)$ and $\Delta_B \in \mathrm{Sp}(n_B)$. The symmetric matrices

$$\Sigma_A = \frac{\hbar}{2}\Delta_A^2, \quad \Sigma_B = \frac{\hbar}{2}\Delta_B^2 \tag{17.25}$$

trivially satisfy

$$\Sigma_A + \frac{i\hbar}{2}J_A \geq 0, \quad \Sigma_B + \frac{i\hbar}{2}J_B \geq 0. \tag{17.26}$$

In view of (16.17), we have

$$\Sigma_A \oplus \Sigma_B \leq \Sigma_U, \tag{17.27}$$

and the theorem now follows from the Werner–Wolf separability conditions. □

17.4 The case of Gaussians

Let us consider again Gaussian distributions of the type

$$\rho(z) = \frac{1}{(2\pi)^n \sqrt{\det \Sigma}} e^{-\frac{1}{2}\Sigma^{-1}(z-\bar{z})^2}$$

where Σ is a positive definite symmetric (real) $2n \times 2n$ matrix. It is the η-Wigner distribution of a density matrix if and only if Σ satisfies the condition

$$\Sigma + \frac{i\eta}{2}J \geq 0 \Longleftrightarrow |\eta| \leq 2\lambda_{\min} \tag{17.28}$$

where $\lambda_{\min}$ is the smallest symplectic eigenvalue of the covariance matrix Σ, as follows from the discussions in previous chapters. In fact, if $\eta > 0$, this is just condition 13.17 in Proposition 168 replacing $\hbar$ with η; if $\eta < 0$, it suffices to note that the condition $\Sigma + \frac{i\eta}{2}J \geq 0$ is equivalent, by complex conjugation, to $\Sigma + \frac{i\eta}{2}J \leq 0$. The purity of the corresponding η-density matrix is

$$\mu(\widehat{\rho}_\eta) = \left(\frac{\eta}{2}\right)^n \det(\Sigma^{-1/2}), \tag{17.29}$$

hence $\widehat{\rho}_\eta$ represents a pure state if and only if $\det(\Sigma) = (\eta/2)^n$.

Proposition 227. *Assume that $\widehat{\rho}$ (i. e. $\widehat{\rho}_\eta = \widehat{\rho}_\hbar$) represents a pure state. Then, for every $\eta < \hbar$, the density operator $\widehat{\rho}_\eta$ represents a mixed Gaussian state.*

Proof. This is clear in view of the previous discussion, using Proposition 47, since the quantum condition $\Sigma + \frac{i\eta}{2}J \geq 0$ is not only necessary but also sufficient for $\widehat{\rho}_\eta$ to be a density operator. □

To summarize, we have the following situation (we assume here for simplicity that $\eta > 0$): Suppose that the quantum condition on the covariance matrix holds for $\eta = \hbar$. Then the system is a mixed quantum state for all $\eta \leq \hbar$; when $\hbar \leq \eta \leq 2\lambda^{\sigma}_{\min}$, it is still a mixed state unless $\eta = \lambda_1 = \cdots = \lambda_n$, in which case it becomes a pure Gaussian state; when $\eta > 2\lambda_{\min}$, we are in the presence of a classical Gaussian state.

17.5 Comments and references

There is a vast literature on Gaussian quantum states and their separability. This is due to the fact that Gaussian states are the easiest to manufacture "in the lab" and also because they are the only states for which separability and entanglement are relatively well understood from a theoretical point of view.

Proposition 216 goes back to the paper [74] by Werner and Wolf where the authors elaborate on earlier work of Werner [73]. For a different proof, see Serafini [65], p. 178.

Proposition 218 is a linear improvement of the symplectic non-squeezing theorem of Gromov [41] in the linear case and which refines a recent previous result of Abbondandolo and his collaborators [1, 2].

Proposition 226 extends to the multipartite case: Adapting the proof, it is easy to show that every multipartite Gaussian state $\widehat{\rho}$ can be disentangled by a symplectic rotation. See de Gosson [34].

Bibliography

[1] A. Abbondandolo, R. Matveyev, How Large Is the Shadow of a Symplectic Ball? J. Topol. Anal. 5(01), 87–119 (2013)

[2] A. Abbondandolo, P. Majer, A Non-Squeezing Theorem for Convex Symplectic Images of the Hilbert Ball. Calc. Var. 54, 1469–1506 (2015)

[3] F. Agorram, A. Benkhadra, A. El Hamyani, A. Ghanmi, Complex Hermite Functions as Fourier–Wigner Transform. Integral Transforms Spec. Funct. 27(2), 94–100 (2015)

[4] Arvind, B. Dutta, N. Mukunda, R. Simon, The Real Symplectic Groups in Quantum Mechanics and Optics. Pramana J. Phys. 45(6), 471–497 (1995)

[5] K. M. Ball, Ellipsoids of Maximal Volume in Convex Bodies. Geom. Dedic. 41(2), 241–250 (1992)

[6] P. Blanchard, E. Brüning, Mathematical Methods in Physics, Distributions, Hilbert Space Operators, and Variational Methods. Progress in Mathematical Physics, vol. 26 (Birkhäuser, 2015)

[7] P. Boggiatto, G. De Donno, A. Oliaro, Generalized Spectrograms and τ-Wigner Transforms. CUBO 12(3), 171–185 (2010)

[8] P. Boggiatto, G. De Donno, A. Oliaro, Time-Frequency Representations of Wigner Type and Pseudo-Differential Operators. Trans. Am. Math. Soc. 362(9), 4955–4981 (2010)

[9] P. Boggiatto, B. Kien Cuong, G. De Donno, A. Oliaro, Weighted Integrals of Wigner Representations. J. Pseudo-Differ. Oper. Appl. (2010)

[10] P. Boggiatto, G. Donno, A. Oliaro, Hudson's Theorem for τ-Wigner Transforms. Bull. Lond. Math. Soc. 45(6), 1131–1147 (2013)

[11] M. Born, P. Jordan, Zur Quantenmechanik. Z. Phys. 34, 858–888 (1925)

[12] M. Born, W. Heisenberg, P. Jordan, Zur Quantenmechanik II. Z. Phys. 35, 557–615 (1925). English translation in: M. Jammer. The Conceptual Development of Quantum Mechanics (New York: McGraw-Hill, 1966), 2nd ed. (New York: American Institute of Physics, 1989)

[13] T. Bröcker, R. F. Werner, Mixed States with Positive Wigner Functions. J. Math. Phys. 36(1), 62–75 (1995)

[14] L. Cohen, Generalized Phase-Space Distribution Functions. J. Math. Phys. 7(5), 781–786 (1966)

[15] L. Cohen, The Weyl Operator and its Generalization. Pseudo-Differential Operators (Birkhäuser, Basel, 2013).

[16] E. Cordero, M. de Gosson, F. Nicola, On the Invertibility of Born–Jordan Quantization. J. Math. Pures Appl. 05(4), 537–557 (2016)

[17] E. Cordero, M. de Gosson, F. Nicola, On the Positivity of Trace Class Operators. Adv. Theor. Math. Phys. 23(8), 2061–2091 (2019)

[18] N. C. Dias, J. N. Prata, The Narcowich-Wigner Spectrum of a Pure State. Rep. Math. Phys. 63(1), 43–54 (2009)

[19] N. Dias, M. de Gosson, J. Prata, Maximal Covariance Group of Wigner Transforms and Pseudo-Differential Operators. Proc. Am. Math. Soc. 142(9), 3183–3192 (2014)

[20] N. Dias, M. de Gosson, J. Prata, Partial Traces and the Geometry of Entanglement; Sufficient Conditions for the Separability of Gaussian States. arXiv:2003.13190v1 (2020)

[21] P. A. M. Dirac, Quantised Singularities in the Electromagnetic Field. Proc. R. Soc. A 133, 60–72 (1931)

[22] H. B. Domingo, E. A. Galapon, Generalized Weyl Transform for Operator Ordering: Polynomial Functions in Phase Space. J. Math. Phys. 56, 022104 (2015)

[23] J. Du, M. W. Wong, A Trace Formula for Weyl Transforms. Approx. Theory Appl. (N. S.) 16(1), 41–45 (2000)

[24] A. Einstein, B. Podolsky, N. Rosen, Can Quantum-Mechanical Description of Physical Reality Be Considered Complete? Phys. Rev. 47, 777 (1935)

https://doi.org/10.1515/9783110722772-018

[25] H. G. Feichtinger, Un espace de Banach de distributions tempérées sur les groupes localement compact abéliens. C. R. Acad. Sci. Paris, Sér. A–B 290(17), A791–A794 (1980)

[26] H. G. Feichtinger, K. Gröchenig, Gabor Frames and Time-Frequency Analysis of Distributions. J. Funct. Anal. 146(2), 464–495 (1997)

[27] H. G. Feichtinger, F. Luef, E. Cordero, Banach Gelfand Triples for Gabor Analysis, in Pseudo-Differential Operators, Quantization and Signals. Lecture Notes in Mathematics, vol. 1949 (2008)

[28] G. B. Folland, Harmonic Analysis in Phase space. Annals of Mathematics studies (Princeton University Press, Princeton, N.J., 1989)

[29] M. de Gosson, Maslov Classes, Metaplectic Representation and Lagrangian Quantization. Research Notes in Mathematics, vol. 95 (Wiley–VCH, Berlin, 1997)

[30] M. de Gosson, Symplectic Geometry and Quantum Mechanics. Operator Theory: Advances and Applications (subseries: Advances in Partial Differential Equations), vol. 166 (Birkhäuser, Basel, 2006)

[31] M. de Gosson, Symplectic Methods in Harmonic Analysis and in Mathematical Physics (Birkhäuser, 2011)

[32] M. de Gosson, The Wigner Transform. Advanced Textbooks in Mathematics (World Scientific, 2017)

[33] M. de Gosson, Introduction to Born–Jordan Quantization. Fundamental Theories of Physics (Springer-Verlag, 2016)

[34] M. de Gosson, On the Disentanglement of Gaussian Quantum States by Symplectic Rotations. C.R. Acad. Sci. Paris 358(4), 459–462 (2020)

[35] M. de Gosson, Quantum Polar Duality and the Symplectic Camel: a New Geometric Approach to Quantization. arXiv:2009.10678v3 [quant-ph] (2020)

[36] M. de Gosson, F. Luef, Remarks on the Fact that the Uncertainty Principle does not Characterize the Quantum State. Phys. Lett. A 364, 453–457 (2007)

[37] M. de Gosson, F. Luef, Symplectic Capacities and the Geometry of Uncertainty: the Irruption of Symplectic Topology in Classical and Quantum Mechanics. Phys. Rep. 484, 131–179 (2009)

[38] M. de Gosson, F. Nicola, Born–Jordan Pseudodifferential Operators and the Dirac Correspondence: Beyond the Groenewold–van Hove Theorem. Bull. Sci. Math. 144, 64–81 (2018)

[39] I. S. Gradshteyn, I. M. Ryzhik, Table of Integrals, Series, and Products (Academic Press, 2014)

[40] K. Gröchenig, Foundations of Time-Frequency Analysis (Birkhäuser, Boston, 2000)

[41] M. Gromov, Pseudoholomorphic Curves in Symplectic Manifolds. Invent. Math. 82, 307–347 (1985)

[42] A. Grossmann, Parity Operators and Quantization of δ-Functions. Commun. Math. Phys. 48, 191–193 (1976)

[43] M. Horodecki, P. Horodecki, R. Horodecki, Separability of Mixed States: Necessary and Sufficient Conditions. Phys. Lett. A 223, 1–8 (1996)

[44] M. Horodecki, P. Horodecki, R. Horodecki, Separability of n-Particle Mixed States: Necessary and Sufficient Conditions in Terms of Linear Maps. Phys. Lett. A 283(1–2), 1–7 (2001)

[45] L. Hörmander, The Weyl Calculus of Pseudo-Differential Operators. Commun. Pure Appl. Math. 32, 359–443 (1979)

[46] L. Hörmander, The Analysis of Linear Partial Differential Operators III (Springer–Verlag, Berlin, 1985)

[47] R. L. Hudson, When is the Wigner Quasi-Probability Density Non-Negative? Rep. Math. Phys. 6, 249–252 (1974)

[48] M. S. Jakobsen, On a (No Longer) New Segal Algebra: A Review of the Feichtinger Algebra. J. Fourier Anal. Appl. 24(6), 1579–1660 (2018)

[49] A. J. E. M. Janssen, A Note on Hudson's Theorem about Functions with Nonnegative Wigner Distributions. SIAM J. Math. Anal. 15(1), 170–176 (1984)
[50] D. Kastler, The C^*-Algebras of a Free Boson Field. Commun. Math. Phys. 1, 14–48 (1965)
[51] Y. Katznelson, An Introduction to Harmonic Analysis (Dover, New York, 1976)
[52] L. Lami, A. Serafini, G. Adesso, Gaussian Entanglement Revisited. New J. Phys. 20, 023030 (2018)
[53] J. M. Leinaas, J. Myrheim, E. Ovrum, Geometrical Aspects of Entanglement. Phys. Rev. A 74(1), 012313 (2006)
[54] J. Leray, Lagrangian Analysis and Quantum Mechanics, a mathematical structure related to asymptotic expansions and the Maslov index (MIT Press, Cambridge, Mass, 1981); translated from Analyse Lagrangienne RCP 25, Strasbourg Collège de France, (1976–1977)
[55] G. Loupias, S. Miracle-Sole, C^*-Algèbres des systèmes canoniques, I. Commun. math. Phys. 2, 31–48
[56] G. Loupias, S. Miracle-Sole, C^*-Algèbres des systèmes canoniques, II. Ann. Inst. Henri Poincaré 6(1), 39–58 (1967)
[57] B. D. MacCluer, Elementary Functional Analysis. Graduate Texts in Mathematics, vol. 253 (Springer, 2009)
[58] F. J. Narcowich, Conditions for the Convolution of Two Wigner Distributions to Be Itself a Wigner Distribution. J. Math. Phys. 29(9), 2036–2041 (1988)
[59] F. J. Narcowich, Distributions of $\hbar$-Positive Type and Applications. J. Math. Phys. 30(11), 2565–2573 (1989)
[60] F. J. Narcowich, Geometry and Uncertainty. J. Math. Phys. 31(2), 354–364 (1990)
[61] T. Paul, A propos du formalisme mathématique de la Mécanique Quantique. Logique & Interaction: Géométrie de la cognition, Actes du colloque et école thématique du CNRS" Logique, Sciences, Philosophie" a Cerisy" (Hermann, 2009)
[62] A. Peres, Separability Criterion for Density Matrices. Phys. Rev. Lett. 77, 1413–1415 (1996)
[63] H. J. Reiter, Metaplectic Groups and Segal Algebras. Lect. Notes in Mathematics (Springer, 1989)
[64] A. Royer, Wigner Functions as the Expectation Value of a Parity Operator. Phys. Rev. A 15, 449–450 (1977)
[65] A. Serafini, Quantum Continuous Variables: A Primer of Theoretical Methods (CRC Press, 2017)
[66] M. A. Shubin, Pseudodifferential Operators and Spectral Theory (Springer-Verlag, 1987) [original Russian edition in Nauka, Moskva, 1978]
[67] F. Soto, P. Claverie, When Is the Wigner Function of Multidimensional Systems Nonnegative? J. Math. Phys. 24(1), 97–100 (1983)
[68] E. M. Stein, Harmonic Analysis: Real Variable Methods, Orthogonality, and Oscillatory Integrals (Princeton University Press, 1993)
[69] E. M. Stein, G. Weiss, Fourier Analysis on Euclidean Spaces (Princeton University Press, 1971)
[70] J. Toft, Hudson's Theorem and Rank One Operators in Weyl Calculus. Pseudo-Differential Operators and Related Topics (Birkhäuser, Basel, 2006), pp. 153–159
[71] V. S. Varadarajan, Lie Groups, Lie Algebras, and Their Representations (Prentice Hall, 1974) (Springer Verlag, 1984)
[72] N. Wallach, Lie Groups: History, Frontiers and Applications. Symplectic Geometry and Fourier Analysis, vol. 5 (Math Sci Press, Brookline, MA, 1977)
[73] R. Werner, Quantum Harmonic Analysis on Phase Space. J. Math. Phys. 25(5), 1404–1411 (1984)
[74] R. F. Werner, M. M. Wolf, Bound Entangled Gaussian States. Phys. Rev. Lett. 86(16), 3658 (2001)
[75] E. Wigner, On the Quantum Correction for Thermodynamic Equilibrium. Phys. Rev. 40, 749–759 (1932)
[76] M. W. Wong, Weyl Transforms (Springer, 1998)
[77] F. Zhang, The Schur Complement and Its Applications (Springer, Berlin, 2005)

Index

ħ-positive type 158

adjoint of a Weyl operator 67
auto-ambiguity function 41
axioms of quantum mechanics 14

Bochner theorem 157
Born–Jordan operator 83
Born–Jordan ordering 83
Born's rule 14

ChaDensity 167
Cohen class 73
compact operator 5
composition of Weyl operators 69
covariance ellipsoid 186
covariance matrix 175
cross-ambiguity function 41
cross-Wigner transform 33

disentanglement of Gaussian states 210

entanglement 193
expectation value 175

Feichtinger algebra 119
Feichtinger state 181, 183
Fermi's trick 51
Fourier transform 8

Gaussian quantum state 205
generators of the metaplectic group 96
generators of the symplectic group 93
Gromov width 185
Gromov's symplectic non-squeezing theorem 184

Hadamard product of two matrices 159
Heisenberg
– group 30
Hermite function 53
Hermite polynomial 53
Hermite's differential equation 53
Hilbert space 3
Hilbert–Schmidt norm 136
Hilbert–Schmidt operator 135
Hudson's theorem 36

kernel of a Weyl operator 66
KLM conditions 158

Laguerre function 54
Laguerre polynomial 54, 56
large numbers hypothesis 19

metaplectic group 96
mixed quantum state 14, 167
modulation space $M^1_s(\mathbb{R}^n)$ 131
Moyal identity 38, 75

Narcowich–Wigner spectrum 179
non-injectivity of Born–Jordan quantization 90

Parseval's identity 4
partial trace 193
PPT theorem 197
pre-Iwasawa factorization 107
pure quanum state 14
purity of a quantum state 171
purity of Gaussian states 205

quantum blob 186
quantum Bochner theorem 157, 159
quantum condition on the covariance matrix 176
quantum observable 14

reduced density operator 193, 195
reflection operator 25
Robertson–Schrödinger inequalities 182
Rodrigue's formula 53

Schur complement 199
Schur product of two matrices 159
separability 193
Shubin pseudodifferential operator 88
Shubin symbol classes 68
Sobolev space 3
spectral theorem 6
statistical interpretation of the Wigner transform 46
superposition of quantum states 14
surjectivity of Born–Jordan quantization 92
symplectic ball 186
symplectic covariance 103
symplectic covariance of the ambiguity function 105

symplectic covariance of the Wigner transform 105
symplectic covariance of Weyl operators 109
symplectic form 10
symplectic Fourier transform 27
symplectic group 10

tempered distribution 7
trace class operators 145
trace norm 152
transpose of a Weyl operator 67

uncertainty and Gromov's width 187
uncertainty principle 181

variable $\hbar$ 18

Werner and Wolf condition 201
Weyl operator 61
Weyl transform 61
Wigner distribution of a mixed stae 167
Wigner transforms of Gaussians 49
Wigner transforms of Hermite functions 53

Advances in Analysis and Geometry

Volume 3
Alexander Grigor'yan, Yuhua Sun
Analysis and Partial Differential Equations on Manifolds, Fractals and Graphs, 2021
ISBN 978-3-11-070063-3, e-ISBN 978-3-11-070076-3, e-ISBN (ePUB) 978-3-11-070085-5

Volume 2
Richard M. Aron, Eva A. Gallardo Gutiérrez, Miguel Martin, Dmitry Ryabogin, Ilya M. Spitkovsky, Artem Zvavitch (Eds.)
The Mathematical Legacy of Victor Lomonosov. Operator Theory, 2020
ISBN 978-3-11-065339-7, e-ISBN 978-3-11-065675-6, e-ISBN (ePUB) 978-3-11-065346-5

Volume 1
Jie Xiao
$Q\alpha$ Analysis on Euclidean Spaces, 2019
ISBN 978-3-11-060112-1, e-ISBN 978-3-11-060028-5, e-ISBN (ePUB) 978-3-11-060010-0

www.degruyter.com